Bigfoot

Surprising Encounters With Bigfoot in
the United States

(A Collection of Unsettling Encounters)

Ursula Vega

Published By **Darby Connor**

Ursula Vega

Bigfoot: Surprising Encounters With Bigfoot in the United States (A Collection of Unsettling Encounters)

ISBN 978-1-77485-602-4

No part of this guidebook shall be reproduced in any form without permission in writing from the publisher except in the case of brief quotations embodied in critical articles or reviews.

Legal & Disclaimer

The information contained in this ebook is not designed to replace or take the place of any form of medicine or professional medical advice. The information in this ebook has been provided for educational & entertainment purposes only.

The information contained in this book has been compiled from sources deemed reliable, and it is accurate to the best of the Author's knowledge; however, the Author cannot guarantee its accuracy and validity and cannot be held liable for any errors or omissions. Changes are periodically made to this book. You must consult your doctor or get professional medical advice before using any of the suggested remedies, techniques, or information in this book.

Upon using the information contained in this book, you agree to hold harmless the Author from and against any damages, costs, and expenses, including any legal fees potentially resulting from the application of any

Table of contents

Introduction

Bigfoot, also called Sasquatch, is thought to be a large, hairy, ape-like being that inhabits some of the dense forests and remote mountain ranges of North America. Although solid, scientific evidence of the creature is lacking, stories of sightings and encounters have been steadily accumulating throughout the years.

As a result, Bigfoot has been placed in the category of so-called "cryptids" — creatures that are thought by some to exist on Earth, but have yet to be properly documented by modern science. Others in this category include the mysterious, waterborne, dinosaur-like being popularly known as "the Loch Ness monster", as well as the goat-eating Chupacabra of South America.

Due to the sheer volume of eyewitness accounts involving Bigfoot, most people in the Western hemisphere recognize the general description of this elusive, ape-like creature. What few may realize, however, is that the Sasquatch is not unique to North

America. Or, at least, stories of mysterious, bipedal hominids are not.

From the mountainous regions of Nepal to the wild outback of Australia, tales of such beings have existed throughout history. In many cases, the descriptions of these creatures seem to match Bigfoot.

So, then, a few questions naturally arise:

Are these all part of a singular, long-lost species of great ape, or are they separate beings with different genetic lineages?

How similar are they to each other, where do they to live, and is there any evidence of their existence?

In this book, we will explore these questions, and many more, as we go through fascinating accounts of sightings, encounters and possible findings relating to the many mysterious ape-men of our world.

Chapter 1: The Case For Hidden Hominids

Before we begin our exploration of Bigfoot and the others, let us first tackle one of the most common objections you will hear when discussing the possible existence of these beings.

By now, there has been several hundred years of reported sightings and encounters. Why, then, is there still no solid evidence of Bigfoot or the other, ape-like cryptids that supposedly live on our planet?

Cryptozoologists are generally hopeful that the advent of easily-accessible, high-definition cameras will lead to clear footage of the elusive beings. Many skeptics however, believe that if there has not been produced anything yet, it is likely comprised of nothing but myths and hoaxes.

However, those with such attitudes would do well to remember the many animals that were once thought of as nothing but fantasy. Many species, which are commonly included in text books today, were not even thought to exist just a handful of generations ago. In regards to Bigfoot, the most relevant example of such an animal is the gorilla.

Since Ancient times, European travelers would return home from Africa with stories of huge, hairy, man-like beasts with incredible strength. The very first historical reference to gorillas is thought to come from Greek texts that documented the life of Hanno the Navigator — a famous Carthaginian explorer. He described them as "savage people covered in hair", which his interpreters called "gorillae" — this is where the common name originates from.

In 1589, an English adventurer named Andrew Battel was sailing with one Captain Abraham Cocke towards the Río de la Plata — a partially-closed coastal body where the Uruguay and Parana rivers meet. Before they could make it all the way there, however, they were forced to return to the Brazilian coastline due extreme wind and hunger.

After desperately docking at São Sebastião, what is today Rio De Janeiro, Andrew Battel and five other men were separated from the rest of the ship's crew. In their vulnerable state, a group of Brazilian Natives captured them and handed them over to the Portuguese who ruled over the region.

After this incident, Battel would spend almost a decade of his life either being imprisoned by, or in service of, the Portuguese. During his time spent traveling with his captors, he would visit several areas of West Africa, where he witnessed gorillas for the first time in his life.

He described them as powerful, human-like monsters that were covered in hair, expect on their faces and ears. He also noted that they had huge fangs, but did not eat any flesh — only fruits, nuts and other plant foods. Despite Battel's stories of the gorilla, however, the great ape remained largely a mystery for Westerners, and scientists believed the animal to be nothing more than a myth conjured up by African natives.

It was not before 1847 that hard evidence finally surfaced, when the American physician and missionary Thomas Savage sent several gorilla bones, including a skull, from Liberia to a colleague back home. By late summer of that same year, the bones had been carefully examined by several professionals.

Together with Savage's notes on the animals' behavior, the gorilla bones allowed him and the Harvard anatomist Jeffries Wyman to publish a scientific paper officially acknowledging the creature's existence. This would prove to be one of the most important discoveries of the latter part of the century, since the findings played a key role in the soon-to-be heated debate regarding Darwinian evolutionary theory.

Another animal once thought of as mythical is the giant panda of China. Today, it is one of the most iconic creatures on the face of the Earth. In previous centuries, however, they had more or less the status of Bigfoot — few people had seen them, and tales of the animals' existence sounded ridiculous to most.

Giant, docile, black and white bears that ate nothing but vegetation? 'Nonsense', was the general consensus of Western scientists. And, for a long time, it looked to be just that. Besides some rare reports of sightings and the odd story from locals, no hard evidence was found until 1869, when a French missionary named Armand David sent a skin of a dead panda back to Europe.

Still, with expeditions actively searching for pandas in the dense bamboo forests of China, nobody managed to find a live specimen to observe or capture. It was not until 1916 that a German zoologist named

Hugo Weigold managed to locate and purchase a cub from a local while traveling throughout China. He became the first European in history to witness a living panda.

20 years later, in 1936, the American fashion designer and socialite, Ruth Elizabeth Harkness, along with a team of trained professionals, managed to find and capture a nine-week old panda cub in the Chinese wilderness, which they named Su Lin. Su Lin became the first living panda to be sent to a Western country, and spent the rest of her life in the Brookfield Zoo in Chicago.

An interesting detail is that not only were pandas completely unknown outside of China for the past centuries, but even in the natives' own documents and artistic works, there is little mention of the animals. In fact, Chinese artists have been regularly depicting bears and bamboo forests for thousands of years, yet no panda was clearly illustrated until the 20th century.

Other notable mentions of supposedly-mythical creatures that turned out to be real include: giant squids, pythons, Komodo dragons, and even giraffes. With so many animals remaining obscure for centuries, it is certainly not unreasonable to think that one or more intelligent species of ape-like beings are living secluded in the deep forests and underground areas of our planet.

Chapter 2: Bigfoot

Bigfoot, also known as Sasquatch, is generally described as a big, muscular, bipedal, ape-like creature. It is almost fully covered in black, dark brown, or dark-reddish hair, expect on its face, hands and feet. It stands between 6 to 9 feet in height, which places it among the larger of the hominid cryptids that have been reportedly seen around the world.

Those who claim they have seen Bigfoot describe a low-set, large forehead, pronounced brow ridge, and big, dark eyes. The top of his head is crested and round, similar to that of a gorilla. The large footprints that give the being its popular name have been described as being as long as 24 inches and as wide as 8 inches. Many who have had encounters with him remark on a very strong, musky and unpleasant odor that is impossible to miss.

Bigfoot is usually spotted in forested areas — often in the Pacific Northwest region of the United States and Canada. Though, he has also been seen in the Great Lakes region of the U.S. Based on the number of sightings reported, it has been estimated by some researchers that between 1500 to 4000 Sasquatch live somewhere in the Pacific Northwest.

Those who have witnessed the beings believe that they are mainly nocturnal, since most sightings and encounters have occurred at night. Based on witness reports, they are also believed to be omnivorous — targeting plants as well as smaller animals for food.

It is not known how Bigfoot communicates, except that they occasionally howl with incredible power — emitting sounds like "ook-ook-ook" and "sooka-sooka" over long distances. In a few cases, they have reportedly slammed rocks and other objects into trees, creating loud thuds at regular intervals, as if they were signaling to each other — or, perhaps, trying to ward off human intruders.

Reports of North American Bigfoot sightings date all the way back to the mid 1800's. However, if we include the earliest Sasquatch tales from Native American folklore, they go back even further. Since these stories are based entirely on oral traditions, though, no one has been able to estimate the earliest mention of the creatures.

According to one of the Native American tribes with the richest Bigfoot lore, the Sts'ailes, the ape-like being's extremely-elusive nature is said to come from some sort of paranormal ability. They believe

12

that the Sasquatch can shift between the physical and spiritual realms, and that, because of this, it is nearly impossible to track down or capture it.

One of the earliest, and most prominent, possible mentions of Bigfoot can be found in a book from 1893 called "The Wilderness Hunter", written by none other than the former U.S. president Theodore Roosevelt. In his book, Roosevelt relays the story of a fellow outdoorsman he had met on his travels — a German-born man he simply referred to as Bauman. He was a rugged, self-sufficient individual that had lived all his life out on the frontier — areas of the United States that had still not been fully explored or settled by Europeans at the time.

One night, after a hunting trip, Bauman and a friend of his, who remained unnamed, were returning to their camp. As soon as they arrived, it became clear that someone, or something, had ransacked their belongings. Supplies and gear had been thrown around, their lean-to shelter had been torn down, and large footprints could be seen all around the site.

At first, the men suspected that a bear had gone through the area. However, when examining the footprints with a torch, it became apparent that a bipedal creature of some sort had made them — or, at least, that was what it looked like. At that point, it was too dark for the men to be completely sure.

After they had eaten, they spent some time discussing the nature of the tracks and whether or not they could belong to a person. After agreeing on the fact that they looked non-human, the two men rolled up in their blankets and went to sleep. However, their rest would not last long.

Around midnight, Bauman was suddenly awoken by some noise near the camp and quickly sat up with his eyes peeled. Immediately, he noticed a strong smell — which he described as a "wild-beast odor". Then, he caught a glimpse of a giant, upright figure looming near the opening of the shelter.

Startled, Bauman immediately grabbed his rifle and fired off a shot towards the threatening shadow. The being seemed to have avoided the bullet, and quickly ran off into the woods at incredible speed. Bauman noted that the creature, whatever is was, could be heard crushing shrubs and small trees as it made its way through the terrain — indicating that it was very heavy. Terrified, the two trappers stayed awake for most the night, only dipping in and out of sleep on occasion. They neither saw nor heard any further sign of the humungous beast, figuring the rifle shot had scared it away.

The next day, they set out into the wilds to check on the traps they had laid the day before, and to set up new ones. When they returned to their camp in the evening, they found that, once again, their equipment had been tossed around and their shelter had been destroyed. Just like the last time, they saw large tracks all around the area — seemingly made by a bipedal creature. The uninvited visitor had been at it again.

Not wanting a repeat of the night before, the men lit a campfire and kept guard one at a time while the other slept. Around midnight, what seemed to be the same creature as earlier appeared on the other side of a stream that ran next to the men's resting place.

It stayed there for a while, partially hidden by shrubs and trees. Since the weather was calm, the men could easily hear branches crack underneath the beast's feet as it moved about. Though it did not come near their campfire, the creature gave off a loud sound every now and again, which Bauman described as a grating, long-drawn groan with a sinister tone.

The following morning, the two men decided to leave the area, as it became obvious that the large creature was displeased with their presence, and that it kept a close eye on them. Before they could move on, however, they needed to check the animal traps they had set out the previous day. In light of the situation, they agreed to stick together while doing so. After circling the forest to collect most of the traps, which to their dismay were empty, the men were nearing their basecamp once again.

It was now noon, and the sun was shining bright in the sky. Because of this, they figured it was more or less safe to move about. They agreed that Bauman would go alone to check the beaver traps, which were the only ones that remained. In the meantime, his partner would head back to camp to prepare their packs for travel. And, so, they split up.

Hours later, Bauman had finished with the beaver traps, which, to his disappointment, were also empty. He returned to the camp, shouting at his friend from a distance, telling him to get up so they could leave — but he got no answer. As Bauman went closer to the camp, he could see that the fire had gone out. When he got all the way there, a horrifying scene awaited him. There was his friend, sitting by the trunk of a downed tree, dead. His neck had been viciously broken, and four large fangs looked to have pierced his throat.

When Bauman checked the body of his deceased companion, he noticed that it was still warm. Terrified, he started running as fast as he could, leaving his equipment and gun behind. After fleeing for some time, he finally reached a group of grazing ponies, quickly mounted one of them, and rode it throughout the night until he had left the area completely. And so, the story ends.

It is worth considering that Roosevelt was impressed with the detail and sincerity of Bauman's story, though he noted that, in his opinion, Bauman's German upbringing may have rendered him vulnerable to superstition. Nevertheless, the former president included it in his book as a legitimate, though unexplained, account of something that actually took place. Whether this ordeal was a legitimate Bigfoot encounter or a misinterpreted run-in with a grizzly bear is still debated within the cryptozoology community.

In the latter part of the 20th century, interest in the creature known as Bigfoot grew a great deal. Eventually, it became part of North American popular culture, and was featured in everything from cartoons to figurines. This great spike in interest was started by an article from the True magazine in December of 1959, which described the discovery of mysterious, huge, human-like footprints that were found in the wilderness of California.

One of the most widely-believed pieces of evidence of Bigfoot is the "Patterson-Gimlin film" — a 16-millimeter motion picture that seems to show a large, ape-like creature crossing a forest clearing. The being, nicknamed "Patty", seemed to be around 7 feet tall, had a muscular body almost completely covered in hair, wide, low-set shoulders, and very long arms.

Additionally, it had prominent, pendulous breasts, which indicated that it was an adult female. Its proportions, gait and overall look was decidedly non-human, and matched most eyewitness reports of Bigfoot.

The footage was originally shot in Northern California in 1967 by filmmakers Roger Patterson and Robert Gimlin. Since then, countless groups and individuals have tried to authenticate or debunk it. As new technology became available, the footage was eventually enhanced and stabilized. This revealed certain elements of the film that were previously hard to make out.

One of the things that stuck out the most was the fact that Patty's leg and back muscles can be seen moving underneath her body hair as she walks. This gives the footage a greater sense of authenticity, as it would be extremely hard to fake, especially during the 1960's.

Dmitri Donskoy, formerly Chief of the Dept. of Biomechanics at the USSR Central Institute of Physical Culture, was one of the professionals who examined the Patterson footage. He concluded that the subject in the film was non-human, based on its peculiar gait and seemingly-heavy body weight — which he said would be nearly impossible to replicate. Donskoy noted that the gait looked natural due to its effortless and confident nature, which he described as "neatly expressive".

The American anthropologist Grover Kantz was initially very skeptical of the Patterson film, based on the still photos he saw in a magazine. However, after carefully studying the footage himself, he was intrigued by the realism of Patty's locomotion. He also observed the incredible width and low position of the shoulders, as well as the unique leg and foot motion — which Kantz stated could not have been duplicated by a man wearing a suit. In the end, he concluded that the film showed a real, unknown creature of some sort.

Although certain people, like the costume-seller Philip Morris, said the footage was a hoax which they played some part in making, none of them were able to substantiate their claims. To this day, the Patterson-Gimlin film remains one of the most intriguing pieces of evidence around — not only in the arena of Bigfoot, but paranormal topics in general.

Roger Patterson died of cancer in 1972. Until the end of his days, he insisted that the film was authentic. Robert Gimlin, who mostly shied away from the media, also stood by the fact that the footage was genuine. In 2005, when he finally began appearing at conferences and giving interviews, his story remained unchanged.

Sightings of, and even direct encounters with, creatures thought to be Sasquatch continues to this day. One such encounter reportedly took place in the state of Michigan in November of 2015. In an interview with the Sasquatch Chronicles podcast, a local hunter named Zack told of his unexpected meeting with a giant, ape-like being in the Michigan woods.

It was about 4:30 in the afternoon. Zack was out in the wilds hunting for deer, but had come up empty-handed so far. As it was getting darker, he decided to head towards one of his most trusted hunting spots. He made his way halfway up a very steep hill and hid behind a large stump, aiming his shotgun downwards in preparation for any incoming deer. Unfortunately for him, none showed up.

After some time, he noticed that the area had become strangely quiet. Suddenly, a huge, black, humanoid shape appeared in the corner of his eye. Zack estimated that the creature was about 9 feet tall. It moved in giant strides up the steep hill, seemingly without much effort. The hunter noted that it almost looked like the being was "on a ramp, going up".

The stench was awful, he said — like a mix between rotting meat and a pack of wet dogs. The being's body was covered in thick hair, its arms were very long, and it had an inhuman amount of dense muscle mass. These are all common descriptions of the Sasquatch of North America.

When it reached the top of the hill, the creature was no more than 15 feet away from Zack. Then, it turned its head and made direct eye contact with him. Zack wanted to run, but found himself completely frozen in fear. What he was looking at became burned into his memory. The Sasquatch's face was partially covered in long, thick hair. Its nose was wide and sunken with large nostrils. The creature's eyes were yellowish green and had a slightly-tired look, which gave Zack the impression that it was sick.

After the brief, but intense, eye contact between the two, the Bigfoot quickly turned around and continued upwards —

disappearing into the darkness of the forest. When the encounter had ended, Zack sat by the stump immobilized for several minutes, before finally making a run for his car.

When he reached the safety of the road, he immediately collapsed and vomited profusely — he was left physically and emotionally drained by the entire ordeal. Suffice to say, Zack now believes fully in the existence of Bigfoot, and always stays alert whenever he is out and about in the woods.

Besides the Patterson-Gimlin film and the large number of eyewitness reports, there have been other clues, such as footprint castings, audio recordings of beastly howls, as well as photos and video of what seems to be Bigfoot (though, admittedly, a lot of them are most likely hoaxes or misidentifications). All in all, there is no shortage of circumstantial evidence for the existence of Sasquatch.

With that being said, nobody has managed to record a video of the creature that matches the caliber of the 1967 Patterson-Gimlin film. As a result, the footage remains the most popular piece of Bigfoot evidence even to this day. Will we ever see something as striking as the Patterson film, but captured with modern, high-definition cameras? Only the future will tell. For the time being, however, the giant, enigmatic ape-man of North America remains well-hidden from humanity.

Chapter 3: Yeti

The Yeti, sometimes referred to as the Abominable Snowman, is a hominid cryptid thought to inhabit the mountainous regions of Nepal and Tibet. According to eyewitness reports, its size and shape are very similar, if not identical, to the North American Bigfoot.

The "mountain man", as he is also called, is said to be ape-like in appearance and considerably taller than the average human. To many, he is an integral part of both Tibetan and Nepalese mythology. In the 19th century, the Yeti quickly became a more well-known creature in Western culture after Bigfoot had entered the mass consciousness with a bang.

Not surprisingly, the giant cryptid is generally regarded by mainstream scientists as nothing but a legend, since there is no conclusive evidence of its existence. Among cryptozoological circles, however, there are serious, ongoing investigations and debates about the mysterious being.

Based on eyewitness reports, it is believed to make its dens in hidden caves, high in the mountains — most of which are inaccessible to humans on foot. Like Bigfoot, the Yeti is also said to be a nocturnal creature, since sightings most often occur at night.

The ape-like beast is largely thought to be an omnivore. Though it relies mostly on meat out of necessity, it has also been seen pulling at branches, bushes and shrubs, seemingly looking for plant foods to eat.

The Yeti was sighted long before the 19th century — even pre-Buddhists believed in its existence. Some worshipped the creature as a God of their hunts, or as a "glacier being". Believers in the Bön religion, an offshoot of Tibetan Buddhism, once thought that the Yeti's blood could be used in some religious ceremonies. As such, it is prominently featured in their traditions.

The first documented, alleged Yeti sighting by a Westerner occurred in 1925, by N.A. Tombazi, a Greek photographer, who was part of a geological expedition from England that was traversing the Himalayan Mountains. Near the Zemu Glacier, he saw a large figure walking across the lower slopes of the mountains in the distance.

It was only about 300 yards away from Tombazi, who described it as a distinctly-humanoid being that was walking upright. He reported that the figure would stop now and then to pull at, or completely uproot, various bushes and other growths. Compared to the snow around it, the creature was dark in color, and wore no discernable clothing.

Tombazi himself did not believe in the Yeti, but was greatly puzzled by the strange figure walking about, apparently naked, in the harsh Himalayan weather. Although the size and behavior of the individual were unusual, Tombazi maintained that it must have been a traveling hermit.

A more direct encounter with a Yeti happened in 1938. One Captain Verner D'Auvergne, who was also a museum curator, was traveling alone in the Himalayas when he suddenly went snow blind during an unrelenting snowstorm. With his vision impaired, he quickly became lost in the freezing terrain.

He stated that, as he was nearing death due to hypothermia, a 9-foot tall being that looked like a pre-historic human, covered head to toe in whiteish-grey hair, rescued him. The creature sheltered him in a cave, fed him and looked over him until he regained his strength. When the captain became healthy enough to make sense of the situation and his surroundings, the supposed Yeti had disappeared.

The most famous incident in the history of Yeti research took place in 1951. British mountaineering legend, Eric Shipton, and the medical doctor and researcher, Michael Ward, were exploring the slopes of the Menlung Glacier, near the border between Nepal and Tibet.

They were both experienced adventurers, and had been part of several demanding expeditions up Mount Everest. This time, they found themselves at an altitude of about 20.000 feet, searching for another route up the famous mountain, when they suddenly spotted a series of strange, deep prints in the snow. The men followed the tracks for about a mile before they finally disappeared on ice.

When they examined the tracks closer, they could see that they looked to be made by large, human-like feet, but with a kind of strange, thumb-like appendage

— something between a human and a primate. They were decidedly enormous — around 13 inches across. To demonstrate this, Shipton placed his trusty ice pick next to one of the footprints as a size comparison and proceeded to take a photograph of it.

As opposed to most Bigfoot tracks which tend to be discovered in muddy terrain, these ones were found in hard snow. Because of this, the imprints were distinct and highly detailed. This fact, combined with the authority of Shipton and Ward, caused the media to take interest in the men's story and photographs. As a result, the Yeti was quickly thrust into British popular culture.

Since then, there have been several attempts by various explorers and research groups to locate the Yeti. One of the biggest expeditions were funded by the London Daily Mail in 1954, which set out to locate and bring back a Yeti to England.

They were gone for 15 weeks and were accompanied by trained zoologists, ornithologists, and hundreds of attendants who carried tranquillizer guns and a huge cage with them. Despite their sizable operation, however, they returned home empty-handed. Even so, they did locate some footprints, strange droppings and parts of skin that could not be recognized by experienced analysts.

Sir Edmund Hillary, a mountaineer and explorer from New Zealand, became the first to climb the great Mount Everest in 1953. He told of several instances where he came into contact with a Yeti. On his record climb up the mountain, both his Sherpa guide and he, himself, observed gigantic, humanoid footprints while ascending the peak.

In 1960, Hillary led a much-anticipated Yeti expedition, traveling with the well-known journalist and author, Desmond Doig. Much like the London Daily Mail expedition in 1954, this quest was also sponsored by a famous name — the World Book Encyclopedia. This gave it a sense of legitimacy that most other expeditions lacked.

The crew brought with them special cameras for infrared and time-lapse photos, as well as expensive trip-wire equipment. Even though they stayed in the area for 10 months, they did not find evidence that would convince anyone that the Yeti exists. Sir Edmund did bring back a scalp and two skins, which he thought might be from the back of a Yeti's head. Much to his dismay, however, it was revealed after further study that they were actually from two bears and a goat.

After their expedition ended with disappointing results, Doig and Hillary determined that the Yeti must simply be a legend. At a later date, however, Doig stated that he felt their expedition had been far too clumsy. No Yeti had been spotted — that much was true. But, even with the many people that tagged along, they also did not observe other animals like snow leopards, whose existence is not in dispute.

Westerners are not the only ones that have launched expeditions to locate the Yeti. In 2008, a group of seven Japanese adventurers and researchers went to the Himalayas to look for the cryptid. Their prime goal was a lofty one — to get clear video footage of the elusive beast, in order to prove its existence once and for all.

After a 42-day stay in the freezing mountains, however, they had failed to produce any such evidence. Still, they did not go home completely empty-handed, as they found, and took photographs of, some strange tracks in the snow, which were about eight inches long and looked very human-like.

While skeptics claimed they belonged to different types of bears, the leader of the search party, Yoshiteru Takahashi, insisted that the tracks did not belong to any known animal of the region. The footprints, along with the many accounts from locals he had interviewed, convinced him that the elusive mountain beast is real.

In 2014, a series of studies lead by Bryan Sykes, an Oxford University geneticist, showed that alleged Yeti hairs from the Bhutan and Lahadk regions of the Himalayas contained DNA similar to that of a prehistoric creature reminiscent of a polar bear, which is believed to have existed between 40.000 and 120.000 years ago.

This may show that the Yeti, somehow, is the result of a kind of genetic engineering, or that the mysterious hominid is, in fact, nothing but a long-lost species of bear that occasionally walks on its hind legs. Alternatively, of course, the hairs that were sent to Sykes and his team may not be from a Yeti at all. The cryptid may still be real, but has mostly avoided leaving behind physical evidence for us to find.

The group who examined the unknown hairs, the Oxford-Lausanne Collateral Hominid Project, also studied alleged samples from other ape-like cryptids, including Bigfoot and the Orang Pendek of Indonesia. The research that has been made public so far, however, has proven to be inconclusive.

Chapter 4: Mapinguary

The Mapinguary, sometimes called Maricoxi, are legendary cryptids located in Brazil that are said to resemble great apes, much like their North American "relatives", Sasquatch. They are said to have red fur, and are different than other creatures described in the native mythology. Some theories posit that these beasts have clear hominid features, while others state they are more similar to a sloth, due to the long claws that have been observed in some cases.

The Mapinguary is said to have long, powerful arms, which, together with its enormous claws, can tear even palm trees apart. It has a back that slopes a bit, and it reaches around 7 feet in height. It is completely covered in matted, thick fur, and is sometimes referred to as "blonde beast" in the area of Patagonia.

It is believed that the Mapinguary lives in Bolivia and Brazil, in the deepest regions of the dense rainforest. It is allegedly quite slow, but is still considered to be dangerous since it moves almost silently through thick vegetation with ease.

Some who have seen the Mapinguary say that it emits a frightening shriek and gives off a foul stench. Bullets and arrows seem to be unable to penetrate its hide, which is similar to that of an alligator. As opposed to alligators, however, it is believed that the Mapinguary usually avoids water.

Though most people believe it to be purely mythical, many cryptozoologists are intrigued by the purported sightings of this mysterious being. One of the top theories states that it may be one of the huge, ancient, sloth-like creatures that are known to have roamed Brazil some time during the Pleistocene era. Many modern sightings seem to hint at this being true.

In 1930, a man named Inocencio was adventuring with his friends in Para State, Brazil, when he became separated from the others. When night fell, he climbed a tree so that he would be safe from the many poisonous and predatory animals that live there. During the night, Inocencio heard a type of loud cry he had never heard before. Then, he saw a large, black figure that looked somewhat like a man. Terrified, he instinctually took a shot at it, and first believed he wounded it. Suddenly, however, it turned around and rushed away with great speed.

In 1975, a miner named Mario Pereira de Souza said he saw a Mapinguary in the mining camp where he worked. He said it screamed loudly before running on two legs towards his direction. Curiously, he described the creature as being unsteady on its back feet, and said that it smelled horrible. De Souza was heavily traumatized by the event, and did everything he could to stay away from the rainforest afterwards.

In the 1980's and 90's, an ornithologist and researcher named David Oren managed to locate 50 eyewitnesses who said they had seen a Mapinguary. This group included rubber plantation workers, mine workers and native tribesmen. Seven of those who were interviewed claimed to have gotten shots off at the creature, although no corpse was ever found.

Oren believes that these were giant sloth sightings, and that they prove that the creatures did, in fact, survive the Ice Age after all — though their numbers had likely been heavily reduced. On the other hand, certain Mapinguary sightings describe it in a way that makes it sound similar to the Bigfoot of North America — being hairy, large and bipedal.

Some cryptozoologists believe that both a sloth-like creature and a more traditional, hominid cryptid live somewhere secluded within Brazil. They say that accounts of sightings may have been mixed together under the same name since they share some features, but that they are actually two different types of creatures. Considering how vast and dense the country's wilderness is, this could certainly be possible.

There has been a great deal of effort expended in search of physical evidence of the Mapinguary. Despite of this, there is only the anecdotal kind, at least for now. During his expeditions, David Oren did find a collection of fur and castings of paws. However, much to his dismay, they were all determined to belong to other, known animals. Even though it has successfully eluded him so far, Oren still believes in the creature's existence.

If we delve into history of the region, we can find evidence that the natives of Brazil may have hunted giant sloths as late as 1895. Footprints, skin and dung that seemed to match the creature were found in an Argentine cave that year. The Natives reportedly had trouble killing the Giant

Sloths because of a bony layer of armor which covered their bodies. The skin found in the cave seemed to match this description. A former governor of Argentina, Ramon Lista, also claimed to have seen a giant sloth in this same area.

The Mapinguary is often mentioned in the folklore of South America's Amazon rainforest. It is sometimes said that the Mapinguary can perform basic speech, which makes some people link it to the South American werewolf mythology. Certain legends say that the creature was once human, but, for whatever reason, got stuck in a more primitive, animalistic form. Whatever the true nature of the Mapinguary, sightings are still being reported, albeit rarely, to this day.

Chapter 5: Yeren

Some people call the Yeren a "Chinese Bigfoot", but his appearance is actually different from that of the well-known North American cryptid. The Yeren stands erect, like Bigfoot does, but is reportedly only around 5 to 6.5 feet in height. Its forehead also slopes up from the eyes, as ours do, rather than resembling the typical gorilla-like forehead of the Sasquatch.

Its eyes are said to be deep-set, and the lips protrude. Upturned nostrils and a bulbous nose are also trademarks of the Yeren. Its cheeks are sunken, and its ears are somewhat like a human's, except larger. Its eyes seem to be almost entirely black, and much larger than ours.

The Yeren's body is covered with long, dark brown or reddish hair, except for the areas around the ears and nose. Its legs are relatively close to human proportions, but the arms are markedly elongated — hanging below the knees. Yeren walk comfortably upright with legs apart, and, based on the footprints that have been found, their feet are roughly 12-19 inches in length. A smaller hominid, only around 3 feet tall, that looks a lot like the Yeren has reportedly been spotted in China as well. However, some cryptid researchers note that these could simply be children of the same species.

Yeren are said to have raided farms of pigs and chickens in search for food, so they are believed to be omnivorous. Still, there are no reported incidents of them eating human flesh. Like other hominid cryptids, they choose to run away and hide rather than attack, and seem to be just as afraid of us as we are of them, if not more so.

Yeren have been in Chinese folklore for centuries, oftentimes they are simply referred to as "Wild Men". Like Bigfoot, they are mainly believed to live in caves in forested, mountainous areas. One of the biggest hotspots of sightings is the Hubei province, which lies in the central part of China.

2000 years ago, a statesman-poet who lived during the Warring States period in China referred to Yeren as "mountain ogres". Later, in the times of the Tang Dynasty, a historian described a band of "very hairy men" roaming around in the region of Hubei. A poet in the Ch'ing Dynasty also described a creature that was much like a monkey, but much larger, and decidedly not a monkey.

Since the Yeren has always been a mysterious and elusive creature, various folk tales naturally began springing up in an attempt to explain it. One of the most popular ones goes like this: The Chinese emperor who ordered the building of China's Great Wall used compulsory workers to complete the grand project. To escape the forced labor, some of these workers retreated to the deepest areas of the nearby woods.

Legend says that their descendants eventually became hairy, large and wild, but supposedly retained their ability to speak. When we look over the reports of Yeren sightings and encounters, however, there is no mention of any kind of speech — only the occasional guttural sounds and primal screams.

Wang Zelin, a biologist trained at the Chicago Faculty of Biology, is said to have observed the Chinese ape-man in the year of 1940. His group was traveling between cities when the cars suddenly stopped because the drivers heard gunshots ahead. When they arrived at the scene, it was discovered that a Wild Man had been shot and was lying in the road, dead.

Zelin said that the creature was quite tall compared to the locals — about 6.5 feet. Its body was completely covered with thick, grayish-red hair. When the body of the dead Yeren was rolled over, it was revealed to be a female, with hairy breasts and all. Her appearance was much like the reconstruction of the Chinese Homo erectus — more or less a female "Peking Man".

However, the hair on the dead Yeren was thicker and longer than the recreated model of our old relative. It was also said to have had big, protruding lips. Locals stated that, when alive, it walked briskly and moved about with ease — even up steep hills. It never spoke, but let out a few loud howls now and again. After the incident, Chinese officials arrived at the scene, and nothing was heard about it afterwards.

In 1950, Fan Jingquan, a geologist, had a brief sighting of two Wild Men. He said that they appeared to be a mother and her son. He reportedly saw them in the Shanxi province, in a heavily-forested area that is largely uninhabited by humans.

In 1961, road workers allegedly killed another female Yeren in the Xishuang Banna area. The Chinese Academy of Sciences sent officials to investigate, but the body was gone when they arrived. They concluded, since the height was claimed to be about 4 feet tall, that it must have been a gibbon instead. 20 years after the fact, however, an anonymous journalist involved in the investigation broke his silence to say that the creature that was killed was, in fact, not a gibbon, but an unknown animal with a human-like appearance.

In 1976, a group of Communist Party members had been driving along a highway in Shennongjia — a mountainous, often misty, area that is covered with dense forest. Out of nowhere, they saw a strange, ape-like creature on two legs, with no tail, covered in reddish hair, in their headlights. Scared by the men in the car, the creature allegedly tried to grab a nearby branch in order to pull itself off the road. However, in its panic, it failed, and fell to the ground before quickly rushing off into the woods.

To substantiate their story, over 100 members of the Chinese army, including infiltration teams, scientists and photographers descended on the area. They brought with them hunting dogs, tape recorders and guns with tranquilizing darts. Army personnel spent almost two years in the region, speaking with hundreds of locals.

They also formed large search parties in an effort to track down the unknown being. However, as is usually the case with cryptids, even with large groups of people actively hunting, they were unable to locate it. One of the search parties thought they saw a Yeren in one instance, but an accidental shooting reportedly scared the creature away before they could photograph it.

Over 200 footprints were collected between 1980 and 1985 in various locations of the Shennongjia forest. Most of them were around 19 inches long, and the stride was an average of eight feet. Many eyewitness accounts were recorded, but no Yeren was photographed or captured.

Chinese theories lean towards the belief that the Yeren is related to an extinct primate that supposedly lived 300,000 years ago. Some people also suggest that larger specimen of certain monkeys are the real creatures in sightings of the Wild Men. The animal most often referred to is the golden monkey — a rare species that could possibly be mistaken for a human-like creature by some observers under confusing circumstances, such as heavy mist. However, since this animal has a prominent tail, rarely walks upright, and has footprints that are distinctly monkey-like, many cryptid researchers doubt this theory.

Dr. Jeff Meldrum. Professor of the Department of Anthropology at Idaho State University and avid Bigfoot researcher, spoke with eyewitnesses to Yeren encounters and examined two plaster casts of alleged footprints. He said the prints showed the stride of a creature with a foot physiology that is different from that of humans. So, everything considered, perhaps the Yeren is actually out there, somewhere in the deep, mist-covered forests of China.

Chapter 6: Chuchunaa

The Chuchunaa are sometimes called "Siberian Snowmen". As opposed to most ape-like cryptids, they have been known to walk around with roughly-cut animal skins worn as clothing. The name "Chuchunaa" means fugitives or outcasts. The fact that they apparently wear clothing of a primitive sort leads some researchers to theorize that they are different from creatures like Bigfoot and the Yeti.

The typical Chuchunaa is between 6 and 7 feet tall — shaped much like a human. It has a protruding brow, broad shoulders, and different colorations of hair that cover most of its body.

These creatures are reportedly found in parts of Russia and Siberia where there are extreme weather conditions. In some parts of Siberia, they are called Mulen, which means "bandit". This is due to the fact that they are known to raid barns and other farm buildings at night. This leads people to regard them as carnivores, since they seem to be on a constant search for animals, most likely for food.

More alarmingly, some locals claim that this creature sometimes eats human flesh, which is not usually reported when it comes to hominid cryptids. According to eyewitnesses, the Chuchunaa have not shown any signs of well-developed communication skills.

Sightings of these beings were originally documented in 1928. The government of the Soviet Union sent more than one expedition to the Yana and Indigirka river regions to document people's accounts of the supposed ape-like beasts.

Certain investigators concluded that these particular hominids may be among the last links between today's humans and our more simian-like ancestors. Parts of Siberia are still barren of human life today, so this could potentially be the last refuge for a species that does not wish to be discovered.

One legend from China says that the Chuchunaa lived in the mountainous Russian region of Verkhoyansk — an area known for its extremely low winter temperatures. As the story goes, the Chuchunaa that lived there would catch and slaughter reindeer, eat them, and then wear their skins on their backs. Also, the beings were said to have screamed horribly whenever they came across humans — either in fear or as an effort to scare off the intruders.

On the other side of the same mountain range, people reported a discreet creature about 6 feet tall, which observed humans but apparently did not want to interact with them. These beings were mostly seen by fishermen and hunters, who regularly go into areas of the wilderness that other people tend to avoid.

There are also modern-day stories of encounters with the Chuchunaa. In 2002, a Russian newspaper ran a story with the title "In Search of the Snowman", which told of two reporters tracking the Chuchunaa. In the Barylas district, an animal that could not be identified was caught in a wolf trap earlier that year.

Unfortunately, when the trappers arrived, it was already dead. This creature was described as looking like some sort of primate. Its whole body was covered in fur, except for its face and feet. It also seemed to have a tail, which is unusual for creatures of this type. Some cryptid researchers have noted, however, that this apparent tail might actually have been part of an animal skin, which, as mentioned, Chuchunaa are known to wear on their backs.

The Russian reporter and her photographer wanted to interview the man who had found the body of the strange being. He had, at the time of the incident, been unable to identify it. Later, he said, the corpse suddenly went missing. According to the local government officials, the man had buried the corpse. Still, some researchers claim that the Russian government quickly swooped in and took the creature's body themselves. Why they would do so remains unclear.

The Russian reporter and her co-worker traveled for two hours by plane and 12 by automobile on dirt roads but, when they arrived, the man who found the creature was no longer living in the village. Eventually, however, they managed to locate the father of the man. He stated that his son had found an animal he did not recognize, with a long tail and an unusual, yellowish coat of fur. He made a sketch of the creature after the corpse allegedly disappeared.

The reporter told the father that the officials claimed his son had buried or hidden the animal, but he flat out denied this. Other residents said that the father had been hesitant to speak about his son's extraordinary finding until the rumors were already circulating in the village.

The reporters could not visit the scene of where the creature was discovered since it lies along the Arctic Circle — one of the coldest areas on the entire Earth. Roads and air travel into this region are only available during mild weather. This is thought to be one of the reasons why the Chuchunaa is so rarely observed compared to other hominid cryptids.

Most sightings of Chuchunaa come from hunters, reindeer breeders and the older residents that live in the remote areas of Russia. According to these individuals, the creature can catch wild reindeer by creeping on all fours up to a herd, before it leaps and runs swiftly to take one animal down. As the stories go, when the Chuchunaa catches a reindeer, by grabbing its head or horns, he quickly breaks its neck by shaking it violently. He also sometimes hunts young elk, but does not interact with the largest of the animals.

It appears that, in this particular area, Chuchunaa have hunting grounds in different seasons. They visit the areas of most bitter cold only when the temperatures are moderate. They move to the tundra to follow and hunt the wild deer around springtime, but eventually return to the forest (their preferred location) when the deer herd migrates there in autumn.

To date, there has been no clear-cut physical evidence of the Chuchunaa brought forward for examination. There are plenty of sightings and accounts, but there is no solid proof that backs up their existence. A few photographs have surfaced that have been introduced as evidence, but they cannot be dated or substantiated by any conventional means.

With that being said, in 2005, an archeological group explored the Denisova Cave in Siberia's part of the Altai Mountains. They discovered a fifth phalanx bone in the cave that could point to a new human relative. Curiously, it is different from both Neanderthals and modern humans. Researchers have dubbed this potential hominid, "X-Woman". One team member described the discovery as ground breaking, since they could have found part of a creature that has not even been on the radar of human development until now.

Ian Tattersall, from New York's American Museum of Natural History, explains that the family tree of humans has many branches, and that it is entirely feasible that there are branches out here that we do not know about yet. If the Chuchunaa is one of them remains to be seen.

Chapter 7: Barmanu

The Barmanu of Pakistan and Afghanistan is another cryptid believed to be related to early humans. It is often described as a cross between an ape and a man, and its name is simply translated to "hairy one". Like the Chuchunaa of Russia, it has been spotted wearing animal skins as rudimentary clothing. This rare, but shared, characteristic, combined with their close proximity to each other, leads some researchers to think that Barmanu and Chuchunaa are actually the same species.

The male Barmanu stands a bit over 6 feet in height and are hairier than their female counterparts, who are shorter in stature and have pendulous breasts, much like "Patty" in the Patterson-Gimlin Bigfoot film. Although their walk is very similar to humans, their bodies are reportedly more muscular. Their arms look long for their height, and their feet are inhumanly big.

The ape-like beings seem to dwell in the Karakoram and Hindu Kush mountain ranges, between the Himalayan and Pamir mountains. Sometimes, they have also been spotted in the Shishi Kuh Valley. Based on eyewitness accounts, Barmanu speech only consists of guttural grunting, but it will occasionally emit higher-pitched shrieks when it is hurt or upset. It mostly feeds on leaves, roots, fungi, berries and nuts. It is also believed to eat meat on occasion, although it is not known for stealing livestock like the Chuchunaa.

Tales of the Barmanu go back centuries. In the early 1900's, those in Spanish expeditions into Northern Pakistan first learned about the creature from the locals. These stories came to the attention of a Spanish zoologist named Jordi Magraner. According to those who met him, Magraner had a very serious and scientific way of enquiry. With over a decade of intense research and documentation, he became the first to bring the Barmanu to international attention.

Initially, Magraner traveled frequently in the areas where Barmanu had been sighted, together with a physician, Anne Mallasse, and a third member of the team that remained anonymous. From 1992 through 1994, the three of them chronicled numerous eyewitness reports and personal experiences regarding the Barmanu. They also discovered footprints that appeared like they were made by apes, only larger.

Rahim Shah, an elderly man who lived in the Alai Valley of Kyrgyzstan, explained to Magraner that he had only seen the Barmanu whenever they came down from the mountains to the valley — which only happened during heavy snowfall. He said he had seen them throughout his whole life, and had even witnessed a mother and child holding hands.

His last encounter, he said, occurred in the early morning, just after dawn. He said he was not afraid, since he had seen the same type creature before, and they never acted aggressive towards him. They did not bother him, and he did not bother them. He fondly referred to them as "the big hairy ones", which is what they are often called in Eastern Afghanistan and Northern Pakistan.

In 2002, Jordi Magraner planned to go back to Europe to share his findings about the Barmanu. Before he could do so, however, he was found viciously murdered under mysterious circumstances. According to the local police, Magraner was found lifeless in a small house he was renting in a village near Kailash valley. It appeared as if his throat had been brutally slit open with a knife. No one was ever convicted of the murder, and the motives remain unclear — though, naturally, different theories exist.

Compared to sightings of Bigfoot, encounters with the Barmanu seem to be more commonplace. Shepherds in the mountains have reported seeing the creatures year after year. They describe encounters as more or less routine occurrences that they expect. However, none of the shepherds, or any investigating zoologist for that matter, has ever discovered a cave or other dwellings in which the creatures might live.

The previously-mentioned mountain man of Alai valley, Rahim Shah, is said to be the person who has seen the Barmanu most frequently. However, not even he has found a single one of their caves or dens. He believes there are fewer Barmanu in existence now, since they are seen less often than they were in his younger years.

A 2005 earthquake may be partially responsible for the lowered number of encounters, since it utterly devastated the area where Barmanu were most often sighted. If they are like most wild animals, the ape-folk may have moved to a safer place as soon as they got a sense of the quake.

Today, there is also more of a government presence in this area than there used to be, which may account for fewer Barmanu sightings. Additionally, many people in Non-Governmental organizations (NGOs) regularly explore the mountains today, making the area less appealing for those who wish to stay hidden.

The nomads who bring herds into the valley for summer month grazing have also had fewer encounters with the Barmanu than they used to have. Some are a bit sad, since they feel that "the big hairy ones" have always been a part of the mountain life. Most local shepherds believe that the Barmanu has the same right to live in the area as they do.

Accounts from eyewitnesses in the area are obviously not enough evidence to sway most skeptics. However, due to the sheer number of sightings in such a small area, it seems highly likely that something out of the ordinary roams this mountainous landscape in the East.

Chapter 8: Almas

Almas, or Almasty, is a singular word in Mongolian which simply means "Wild Man" — referring to the mysterious hominid creature that has been spotted in certain areas of the country. This type of being is said to inhabit the Pamir and Caucasus Mountains in Central Asia, as well as the Altai Mountains in southern Mongolia. On rare occasions, they have also been reported in certain parts of Russia.

Like the Cuchunaa and Barmanu, Almas are usually considered closer to being wild, hairy humans than apes, in both their habits and appearance. This contrasts them with cryptid hominids like the Himalayan Yeti, which is believed to be more beast than man.

Almas are generally described as bipedal creatures standing 5 to 6.5 feet in height. Their bodies are covered with fine, brown-reddish hair. Their facial features include a weak chin, prominent brow-ridge and a flat nose. Almas are generally unclothed, although some people have seen them wearing primitive animal skins as clothing, just like the Chuchunaa and Barmanu. It is worth noting that the common eyewitness description of the Almas is very similar to the reconstructed models of the Neanderthals. The only big difference seems to be the abundant, thick body hair of the Mongolian Wild Men.

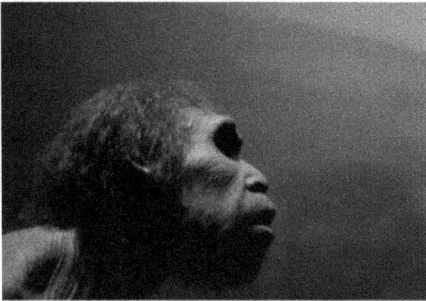

Like most hominoid creatures, the Almas seem to prefer living in forested, hilly or mountainous areas. Unlike most creatures of his type, however, humans living in the surrounding villages describe themselves actually communicating with them, albeit in a primitive way — mostly using hand gestures.

Locals say that the Almas feed mainly on berries and other plant foods, but individual specimen have sometimes been seen hunting animals as well. They are nocturnal, and thus very difficult to spot most of the time. As far as locals know, Almas do not leave any trace of their death in areas frequented by humans.

There is some evidence that suggests that these hominids may be more real than imaginary. This includes many hundreds of eyewitness accounts and findings of strange, human-like footprints. Documented reports of sightings and interactions with Almas go back hundreds of years, and, interestingly enough, the descriptions of the creatures have not changed during that time.

As far back as 1430, Johann Schiltberger, a Bavarian traveler and writer of noble descent, stated that he observed the Almas in Mongolia when he was a captive of a Mongol Khan. When he returned home later in his life, he wrote a book titled "The

Bondage and Travels of Johann Schiltberger", relaying his incredible experiences.

In his memoir, Schiltberger clearly mentions the Almas, which he refers to as savages. He said they would "run about like other wild beasts in the mountain", and that they would eat leaves, grass, and whatever they could find. He mentioned that their bodies were covered with hair, except for their hands and faces.

Nikolai Przhevalsky, a Russian geographer and explorer, conducted five major expeditions in his life time; one to the Russian far east, and four to Mongolia, Zinjang and Tibet. During his travels, he documented a number of animals that were unknown to most people outside of these areas. One such animal was a kind of wild horse, which was the last surviving species that was not widely domesticated by humans. Today, the species is called Przhevalsky horse, named after the Russian explorer.

The horses were not the only new creatures Przhevalsky came upon during his expeditions, however. In 1871, he reported seeing the elusive Wild Men of the mountains himself. Though he never managed to get close enough to study the strange beings, the Russian explorer was quite serious about their existence. He also noted that these beings were listed in the Tibetan and Mongolian apothecary books, alongside many other plants and animals

that are known to exist today.

From 1890 t0 1928, a Russian professor named Tysben Zhamtsarano was purportedly conducting research into the Almas. As was often the case in Russia at that time, the scientist was eventually thrown into a gulag. His illustrations, notes and other evidence, which is believed to have been quite extensive, disappeared along with him. Zhamtsarano's assistant, who avoided the gulag, claimed that he saw much of the evidence himself, including the skin of an Almas that had been preserved in a Buddhist monastery in Mongolia.

In 1963, another Russian, a children's doctor named Ivan Ivlov, reported that he saw an entire family of Almas at a distance of about half a mile. His experience was validated by his personal driver, who also claimed to have observed the Wild Men. Furthermore, being a pediatrician, Ivan Ivlov regularly spoke to Mongolian children, and many of them said they had also seen the strange hominids. Some even claimed to have interacted with them, and stated that the human and Almas children were not afraid of each other.

Local Mongolian legend tells about a "wild woman" named Zana who once lived in a remote mountain village. Some researchers believe that she might have been an Almas, but there was no hard evidence to confirm this. As legend tells it, she was captured by humans in 1850. She was at first quite violent, likely in response to her sudden kidnapping, but eventually calmed down and settled into her new surroundings.

After she got accustomed to the village life, she is said to have had a sexual relationship with a local man who she later married. Together with him, she gave birth to several children who appeared to be more or less human. One of these children were reported to have died during their infancy, allegedly because Zana's genetics were not fully compatible with those of her human partner. According to locals, Zana died in 1890 — though the cause of her death remained unclear.

The story goes on to describe how her husband gave away four surviving children to different families of the village. They were all assimilated into the society and eventually married, each having a family of his or her own. A physician later examined the skull of one of Zana's children after his death and reportedly found it to be a human skull. If this is true, it is possible that Zana was simply from a tribe of hunter-gatherers, rather than a true Almas. Though, if the children were in fact hybrids, the skulls may have looked more or less human anyway.

A French cryptozoologist, Bernard Heuvelmans, and a British anthropologist, Myra Shackley, speculate that Almas are related to the Neanderthals. Another cryptozoologist, Loren Coleman, has suggested that they may be the only surviving descendants of our upright human relatives — homo erectus.

There are some cryptid researchers who insist that the Almas is closely related to the Himalayan Yeti. The locals, however, see themselves as being not much different to the Almas. They see them as simple Wild Men — untouched by the refining qualities of civilization.

There does exist alleged Almas footprint castings and even some debatable photographs that are circling the internet. However, no indisputable physical evidence that the Almas exists has ever reached the mainstream scientific community. It is speculated that, like other cryptids, researchers will need to find a whole body, or even capture a live Almas, in order to acknowledge their existence.

In 2003, the ongoing controversy over whether the Almas is real or not was briefly rekindled. The BBC (British Broadcasting Corporation) reported at that time that Sergey Semenov, a mountain climber, had discovered the leg and foot of a creature he could not identify in the high-altitude permafrost region of the Altai Mountains. X-rays and tests showed that the bones were thousands of years old, but all attempts to identify exactly what creature they belonged to were inconclusive.

Chapter 9: Batutut

The Batutut, or Ujit, of Vietnam is speculated to be either a lost tribe of primitive humans or an undiscovered hominid species like Bigfoot. Based on most eyewitness accounts, it appears to be the latter.

The height of the creatures can vary, from around 5 to over 7 feet tall. Their bodies are almost completely covered in black, brown or gray hair, except on their faces, feet and hands. According to observers, they walk comfortably upright, much like humans. Vietnamese scholars refer to the creature as Ngoui Rung — which loosely translates to "forest man".

The Batutut seem to forage for most of their food, which usually includes leaves and fruit. From time to time, they also kill and eat small animals, like langurs and foxes, though they prefer to eat snails and mussels for meat. In general, these creatures do not seem to bear ill will toward humans. In remote villages in Vietnam, they have even been reported to wander up close to campfires and calmly sit nearby.

According to eyewitness accounts, they do not speak, but, like most hominid cryptids, they do occasionally make some strange noises. They seem to know what each one is saying when seen in small groups. According to observers, they have a very dense musculature, which give the tallest of them an imposing presence.

Dang Nghiem Van, an anthropologist in Hanoi, collects Batutut information from Vietnam's central highlands to the northern parts of the country. Based on what he has uncovered, they are commonly described as extremely powerful yet elusive creatures, and their movement has both human and ape-like characteristics.

The locals believe that Batutut steal people's personal belongings from time to time. They have reportedly even stolen guns, although they have not shown any ability to fire them. As in most countries, areas of uninhabited wilderness are steadily decreasing in Vietnam, which means, for better or worse, that it should be harder for the creatures to remain in hiding.

In the 1960's, American soldiers deployed in Laotian and Vietnamese jungles encountered strange, scary creatures which they had never seen or heard of before. One sergeant named Thomas Jenkins was interviewed after reconnaissance missions in Vietnam. He said that in 1969, he saw a group of what looked like big apes throwing rocks at members of his platoon. This incident gave the Batutut their new, Western, name, "Rock Apes".

Jenkins does not believe that the creatures they saw were normal apes or monkeys, even though biologists say there are many of those in Vietnam. Instead, Jenkins insists that these beings were darker in color, walked upright, and were more muscular than simple monkeys.

The creatures the soldiers observed would sometimes wake them up in the morning, screaming, yelling and shaking their fists. Jenkins said it looked a lot like a human behavior, and that it was obvious to everyone that they objected to the presence of soldiers in the jungle. Unfortunately, since most soldiers in the 1960's did not have cameras with them, no one managed to take a photograph of these unknown hominids.

In addition to American soldiers, North Vietnamese fighters also reported what they called "forest people". They stated that the large creatures attacked their soldiers, too. So, it would seem that the so-called

Rock Apes did not take any side in the human conflict, but rather wanted all of them gone from their territory.

The Bigfoot researcher Cliff Barackman states that his father told him an interesting story of another American soldier who served in Vietnam. One day, the platoon of this soldier was firing on the North Vietnamese in the deep jungle. Suddenly, a mysterious, humanoid creature walked through a small clearing, before climbing a very steep embankment and disappearing back into the jungle. The animal apparently scaled the sharp incline easily, which impressed the American soldier. This account echoes many sightings of Bigfoot, which is also known to effortlessly traverse rough wilderness with inhuman speed.

A retired US forces helicopter pilot, Larry Wilson, allegedly had his own encounter with the Batutut in 1970. He was flying into a valley in the Vietnamese wilderness when he spotted a tree, stripped of its leaves, wiggling fiercely. He then saw a strange ape-man shaking the tree with astonishing power. Wilson described the creature as having a flat skull, roughly the size of a soccer ball, and facial features similar to humans. He had a clear, unobstructed view of it, and was certain it was part ape and part man.

Searching for Batutut in Vietnam's dense jungles is close to impossible these days. There are plenty of unexploded land mines and bombs from the war that make it extremely dangerous for any person traveling on foot. The roads also tend to wash out in the rainy season, which make them nearly impassable. Furthermore, the Vietnam jungle is so incredibly dense that you can hardly see a foot ahead, even when several people are clearing with machetes.

Logistics aside, Josh Gates of Destination Truth (an American paranormal reality TV series) went there in an effort to find the Batutut in 2012. He interviewed Dr. Tran Hong Viet, who works at the Cryptozoic Research Center of Vietnam. This institution was initially set up by the government in order to find and study the elusive forest people.

Dr. Viet has researched the case of the unknown ape-beast for over 30 years and is a strong believer in its existence. He found a footprint some years ago in a very remote cave. The print was not that of any known primate, nor was it human. While in Vietnam, Josh Gates took many photographs of the fascinating print, which was shown on the Destination Truth show.

After talking to Dr. Viet and learning about his findings, Gates pondered where to go in order to search for the Batutut. The vast and uninviting Kebang National Park is barely explored, so he decided to take his crew there. While traversing the terrain, they were filming as they went, documenting nearly every step of their journey.

During their time there, something threw a stone at them out of nowhere. Heavy footsteps were also heard rushing towards their direction before suddenly backing off, at the same time as they heard loud, animalistic calls like they had never heard before. Despite these threatening happenings, however, no Rock Ape made a face-to-face appearance with the crew while they journeyed through the jungle.

The highlight of the whole search was a fresh footprint found by Gates, who carefully cast it and brought it home to the U.S. for additional study. Professor Jeff Meldrum examined Gates' cast and confirmed that it is an authentic sample of an unknown Vietnamese hominoid. These fresh findings give hope to the idea that the mysterious Batutut will eventually be discovered.

Chapter 10: Yowie

In Australia, hairy, hominoid creatures are sometimes interchangeably referred to as either a Yowie or a Yahoo. However, based on eyewitness accounts, these are actually two separate beings with their own unique characteristics.

A Yahoo is described as spending more time on four legs than two. Standing upright, it is about 5 feet in height. It is furry on its sides and back, with lighter-colored fur on its belly, head, arms and legs. Based on sightings, it resembles a thought-to-be extinct type of marsupial that resembled an ape.

The Yowie, on the other hand, is much bigger and stranger looking, and seems to walk comfortably on two legs, like the Yeti and Bigfoot. It appears more human than the Yahoo, but more monster-like as well — due to its huge stature and incredibly-dense musculature. From here on out, we will be speaking mainly about the Yowie.

Yowies are much like other hominid cryptids in that they seem most at home in a forested, mountainous area. In Australia, some of their habitats are flat, while others are hilly. As is the case with similar creatures, people who flee from an encounter with a Yowie have found that when they reach a clearing or road, the creature will stop following them.

On some occasions, Yowies have been seen hunting and killing small pigs and chickens. Most of the time, however, they seem to exist on berries that they pull from branches and on the bark of certain trees, in addition to other plant foods they can find.

Aboriginal folklore has tales of the Yowie that go back for ages, possibly many hundreds of years, if not more. According to the Sydney Morning Herald, the first Western sightings of Yowies occurred in 1795. Back then, they were simply referred to as "indigenous apes".

In 1876, the Australian Town and Country Journal cited Aboriginal beliefs of hairy men who lived in the woods, and who were not animals, yet not quite human. In 1882, another article mentioned the eyewitness account of a naturalist named Henry James McCooey, who claimed to have seen one of these mysterious apes near the southern coast of New South Wales.

As is the case with most cryptids, reports of Yowie sightings were quite rare, but continued to pile up as the years went by —

slowly but surely painting a clear picture of the creature's physical features and habits. In 1982, exactly 100 years after the story of James McCooey was published, another Australian man had an astonishing encounter with a strange, half-human, half-ape being.

The man was walking along the coast of New South Wales when he suddenly heard cries of small birds. When he went to look at what was causing the commotion, he saw that they were darting around a strange-looking creature. The mysterious beast looked to be on its hind legs, but not fully upright, and was looking at the birds that were circling around it. It was blinking its eyes and making a low, chattering noise.

The man was at an elevation higher than the Yowie, so he was easily able to observe its general appearance. He reported that the being would be less than 5 feet in height if it stood upright. It was covered with long hair, which was black, except for reddish patches near the breast and throat. Its eyes were small and partially hidden by some of its matted hair.

The Yowie's arms were slightly longer then its legs — somewhere between a chimpanzee and a human, which is a common indicator of Bigfoot-like beings. The man was initially shocked and repulsed by the creature. Terrified, he threw a rock at the unsuspecting Yowie, which subsequently rushed off — disappearing into a ravine.

As mentioned, this sighting took place in the Australian state of New South Wales. This place has turned out to be somewhat of a hotspot for Yowie encounters, especially near the more mountainous regions, such as the Brindabella Range, which is a popular area for hikers.

In contrast to the North American Sasquatch, the Yowie seems to be more daring when it comes to venturing near humans. While Bigfoot rarely comes anywhere near our territories, Yowies, based on eyewitness accounts, will gladly roam alongside hiking trails or empty cabins. However, that is typically where they stop — they seem to shy away from bright-lit streets, camping grounds and other busy areas.

In 1995, an Australian man named Dean Harrison, formerly a Yowie skeptic, found himself suddenly turned into a believer. He initially figured the stories about the ape-man to be nothing but camp fire tales and jokes for the pub. He states that his skepticism abruptly ended after a series of terrifying encounters, which forced him to accept what mainstream science has not: that Yowies are very much real.

Harrison arrived at his home in Queensland late one night in 1995, with no idea that his life was about to be turned upside down. When he got out of his car, he heard a loud commotion near his back fence. It was a deep growling followed by branches being violently snapped. Harrison went to check it out, before suddenly stopping dead in his tracks. In front of him was a huge, hairy, bipedal beast, crashing around and pulling out small trees from their roots.

Harrison still recalls the hulking creature having a mix of human and ape-like features that he had never seen before. Terrified, he ran into his house and tried to come to terms with what he had just witnessed. Moments later, the Yowie had disappeared.

After the shocking event, Harrison immediately started searching everywhere he could for information about the monster he saw near his backyard. Much to his dismay, he did not find much information available. In fact, almost every mention of the ape-like being was accompanied by a mocking tone — no mainstream organization took it seriously.

Not wanting to appear as a crazy person or a liar, Dean Harrison kept the encounter to himself and tried to move on. However, two years later, in 1997, Harrison would have his second meeting with a Yowie. This time, things would become much more personal.

It was late at night, and Harrison was jogging in a thicket next to the suburban streets of his hometown of Ormeau, Queensland. Before heading further in, he stopped to talk to his wife on the phone. In the middle of the conversation, he suddenly heard strange noises coming from behind him. Foliage was being parted, and something big was steadily heading his way.

Dean Harrison told his wife that he thought it may be a man trying to sneak up behind him. He ended the call, turned towards the sounds and took a defensive position — readying himself for whatever came through the bushes. To his surprise, out came a massive, humanoid silhouette — the Yowie had arrived. It quickly spotted Harrison standing there, and slowly and ominously crouched down between the foliage.

Harrison, consumed by the instinct to flee, made a sudden move with his foot. This provoked the Yowie, which then gave off a bestial roar before lunging forward. Immediately, Harrison got a powerful adrenaline rush, and he started sprinting away as fast as he could.

As he ran along a cleared path, the huge beast ran more or less parallel with him. It effortlessly maneuvered its way through the thick forest — evading logs and rocks like they were nothing. Harrison quickly realized that the Yowie was trying to cut him off.

Fearing his life was about to end, he abruptly changed direction, and managed to buy himself just enough time to reach a well-lit road nearby. Realizing the creature was no longer chasing him, Harrison looked back. The massive beast had retreated into some bushes at the edge of the road, once again squatting down partially hidden while staring right at him.

Dean Harrison recounted his impression of the terrifying creature:

"It was so powerful... If this thing had got a hold of me, I wouldn't have time to scream — it would've snapped my neck like a toothpick..."

This second encounter with Australia's mysterious, hulking ape-man turned Harrison's life upside down. Due to the raw emotional impact, he decided he would now openly discuss and search for the creature.

Later that year he founded the Australian Yowie Research Group, and started the website Yowiehunters.com.au.

Since then, he has made it his passion to collect evidence for, and document sightings of, the Yowie. As he speaks to those who claim to have seen the creature, Harrison asks them to sketch what they saw. He has found that most of the portraits, many of which can be found on his website, are decidedly similar: larger and more muscular than humans, covered in fur that ranges from dark chocolate to reddish brown, with a face that seems to be part man, part ape. This is another reason Harrison believes he saw the same creature they did — the vast majority of the eyewitness descriptions match each other.

Though reports of sightings and close encounters continue to accumulate, hard evidence of the Yowie's existence is still lacking. There are some interesting videos that were allegedly taken of the hulking ape-man, but due to their quality, nothing can be said for certain.

Additionally, there are casts from Australian creek beds that were sent to the United States for examination. Forensic experts confirmed that they are quite similar to the prints of North America's own Bigfoot, which researchers have been casting for years. As is the case with other hominid cryptids, however, these imprints alone will not sway most skeptics.

Dean Harrison notes that since technology is changing at a rapid pace, the chances of getting footage of the Yowie will increase. He hopes that, with the advent of high-definition dash cams and smart phones, clear footage of the massive beast will eventually emerge.

Chapter 11: Orang Pendek

"Orang Pendek" means "short person" in Indonesian. This is the name given to a small, hominid cryptid that is believed to live in the jungles of Sumatra — a large island located in western Indonesia. It is described as an ape-like being that stands 3 to 6 feet tall, with a powerful, humanoid physique that separates it from the rest of the local animal life. It is almost completely covered in short, brown, golden or greyish hairs, and moves around efficiently on two legs.

Its name is indeed reminiscent of the well-known orangutan, which translates to "person of the forest". However, according to eyewitness testimony, the Orang Pendek has facial features that are more similar to those of humans rather than apes or monkeys.

Orang Pendek are reported to be non-aggressive in most cases — it prefers to hide away in the deepest corners of the Sumatran jungle. However, when they have come in contact with humans, they have been known to stretch their arms into the air while making guttural sounds —a common tactic primates use when trying to appear more threatening.

The diet of the Orang Pendek is said to consist mainly of plant foods like young shoots, durian fruit, and ginger roots. Given the opportunity, it will also feed on insects or smaller animals, like crabs. Occasionally, the stocky ape-men have been observed helping themselves to a farmer's harvest.

They seem to particularly enjoy corn, as well as all kinds of fruit.

The Orang Rimba (meaning "people of the forest"), also called Suku Anak Dalam (translated to "Children of the Inner-forest"), is a group of indigenous, animist peoples that live throughout the dense Sumatran forests. According to them, the upright ape-men have been living in the wilderness for centuries. The Orang Rimba elders know where most of the Orang Pendek territories lie, and will occasionally leave behind offerings for them. They do not see the creatures as mysterious or threatening at all. To them, they are simply one of the many co-inhabitants of the forest.

While the tribes of the forest are not typically afraid of the Orang Pendek, the everyday people of the surrounding villages have a different relationship with the hairy beings. They view them as intelligent creatures with a near-supernatural ability to hide from humans. This makes many villagers afraid of the Orang Pendek — they make sure to watch their backs whenever they are walking in what they perceive to be "short people" territory.

From a Western perspective, the Orang Pendek was first heard about in the beginning of the 1900's, when Indonesia was still a Dutch colony referred to as the Dutch East Indies. In 1917, the Dutch Sumatran governor received reports of an unknown hominoid creature living in the dense forests of the island. One of the most peculiar documented cases was that of a Dutch plantation owner simply referred to as Mr. Oostingh.

While he was walking through the forests near the Bukit Kaba mountain in Sumatra, Oostingh suddenly spotted a strange creature sitting on the ground, approximately 30 feet away from him. The being, he said, was about as large as a medium-sized native Indonesian, but with extremely thick, square, shoulders. Also, its body was completely covered in hair, with a dusty, almost grey, blackish color.

Oostingh recalled his experience with the Orang Pendek as follows:

"He obviously noticed my presence. He did not turn his head, but stood up on his feet: he seemed about as tall as I am, about 1.75 meters. Then I realized that this was not a man, and I started back away since I was not armed. The creature took several paces and then, with his long arm, grasped a small tree, which seemed to almost break under his weight, before quietly springing into a tree, swinging in great leaps to the right and to the left.

My main impression was, and still is, 'what a huge beast!' It was not an orangutan — I had seen one of those apes before at the Artis, the Amsterdam Zoo. It was more like a monstrously-large siamang, but a siamang has long hair, and there was no doubt this one had short hair. I did not see its face since, indeed, it never looked at me once."

6 years later, in 1923, while surveying land in Sumatra, another Dutch settler named Van Heerwaden sighted a similar creature of unknown origin. He reportedly saw what looked to be some kind of ape sitting on the branch of a nearby tree. It was unlike anything he had ever seen before.

He described it as dark and extremely hairy, but with facial features that were neither ugly nor ape-like in appearance. Van Heerwaden noted, however, just like Oostingh, that the being had very long arms, which would reach just a little above its knees if it stood up. Also, its legs seemed to be very short. These two factors combined made its body look decidedly non-human. Furthermore, while the settler could not make out the creature's feet, he could see its toes, which he said were quite normal-looking — meaning, they looked much like ours.

Van Heerwaden's detailed observations are typical of most Orang Pendek sightings. Talk about elongated arms can certainly make casual skeptics conclude that they are nothing but orangutans viewed at an obfuscating angle. However, two peculiar things that seem to stick out the most are the feet and faces of these beings. They give onlookers the distinct feeling that they are not witnessing a common ape or monkey. Instead, it regularly conjures up feelings of amazement or fear that no everyday animal can.

When it comes to modern Orang Pendek research, Debbie Martyr, a former English journalist, is the most notable. She spent over two decades doing wildlife conservation work and research in Sumatra, promoted and funded by Fauna and Flora International — a global non-governmental organization.

During her stay in the region, Martyr interviewed hundreds of witnesses who claimed to have seen the Orang Pendek. One of her team members, a professional wildlife photographer named Jeremy Holden, also set up various camera traps in the areas where the ape-man had reportedly been spotted.

Debbie Martyr and her team documented many rare animals in Sumatra, and even rediscovered a lost species of deer — the muntjac, which had not been seen since 1930. Despite their many successful ventures, however, the team never managed to get a clear photograph of the Orang Pendek. Even so, several of them claim to have personally witnessed the rare hominoid at several occasions.

Debbie Martyr reportedly got her first glimpse of the creature in 1990 — only one year after arriving in Sumatra for the first time. She described the being as a relatively small, though immensely powerful-looking, non-human primate. She noted that it did not look like an orangutan, since its proportions did not match. The creature was built much like a boxer, she said, with immense upper-body strength. It had a beautiful color, and moved bipedally in an efficient manner while trying to avoid being seen by the humans.

Jeremy Holden did not have such a direct sighting, but caught a glance of the Orang Pendek while it went over a hill, deep inside the Sumatran forest. He saw the being from behind as it swiftly made its way through the wilderness. Like Martyr, Holden noted that the creature moved upright — more like a human than an ape.

In September of 2001, a team of three independent British explorers and cryptozoologists — Adam Davies, Andrew Sanderson and Keith Townley — traveled to

Sumatra to search for evidence of the Orang Pendek. In the three weeks they were there, they did not see the creature themselves. However, they did find some compelling evidence near Kerinci — a Sumatran volcano surrounded by dense, lush forests.

While they were scouring the area, Davies and his team spotted a set of strange footprints in the muddy ground, which they proceeded to make a cast out of. Nearby, they also found some peculiar, long strands of hair that were yellow-brown in color. Excited, the crew traveled home with their findings. They gave the footprint cast to Cambridge Professor Dr. David Chivers, who compared it to other known Sumatran animals. He deduced that it was definitely from some sort of ape, which seemed to have a mix of characteristics from humans and other known hominids.

Dr. Chivers' final statement on the matter was as follows:

"From further examination, the print did not match any known primate species, and I can conclude that this points towards there being a large, unknown primate in the forests of Sumatra".

The strange hairs that the three men found were sent to an expert on mammal hairs, Dr. Hans Brunner, at Deakin University in Melbourne, Australia. After performing extensive DNA testing, he determined that the hairs found in the Sumatran jungle did not match any of the known local animals, nor did they belong to a human. His final statement was that the hairs seemed to belong to an unidentified type of primate.

At the start of 2005, National Geographic provided funding for a project aimed at getting photographic evidence of the Orang Pendek. It was led by Dr. Peter Tse from Dartmouth College, New Hampshire, and involved placing a number of camera traps in the regions where the cryptids had been commonly sighted. Unfortunately, the project concluded in 2009, after many years of frustration due to lack of success.

Today, despite the intriguing footprints and hair samples found in 2001, there are still no clear photographs or video footage of the mysterious Orang Pendek. However, as eyewitness accounts continue to pile up, many researchers still believe its existence can be proven in the future.

Chapter 12: The Legends Of Bigfoot From Around The World

Whether Bigfoot exists is still up for debate.Some people insist it is only a myth, while others claim so many individuals could not possibly be wrong.Sightings and encounters with Bigfoot have occurred in countries all around the world.

What Is This Legendary Creature?

Virtually everyone who has encountered the creature state it has had some common traits.One example is its physical appearance.The creature is described as hairy and ape-like.A second example is its behavior.It is said to stand straight upright, and move like a human being.

However, the trait that often comes to mind is shared by most but not all of these creatures.Although the majority are described as very tall, there are exceptions.Some are described as human-size, or even smaller.

Bigfoot:The United States

Sightings of Bigfoot have been reported in every state in America.Thus far, Washington State has had the most sightings.The majority of Bigfoot sightings have been in the Pacific Northwest.

Bigfoot is generally described as standing between 6 to 9 feet tall, muscular, and covered with dark hair.Its huge feet are up to 24 inches long.When footprint casts were taken, they showed claw marks.This has led some to believe Bigfoot is a bear.Regardless of its similarity to apes, apes do not have claws.They have fingernails, like humans.

The legend of Bigfoot started long before its name.Initial versions described it as a threatening creature to be feared.Other versions stated they did nothing more menacing than stealing salmon.

Scientists have tried to explain it away with logic, but explanations have not been compatible with the facts.Bigfoot does not share enough characteristics with present-day animals, and other characteristics were only compatible with beings that were only found in other countries or were completely extinct.

Sasquatch:Canada

Sasquatch sightings have occurred in Canada for more than a century.The term itself translates to hairy man or wild man.After an 1864 sighting, creatures were described as hairy humanoids that attacked fur traders with rocks.

Encounters have resulted in differing descriptions of a Sasquatch's behavior.They range from quiet and fearful to curious and aggressive.Most accounts agree that the creature does not like to be around humans.

Whether it is real or not, Canadian legislators introduced a petition to have it declared an endangered species, and for it to be protected by law.

Yeren:China

The Yeren is alternately referred to as the Man Bear, Man Monkey, and Chinese Wildman.

Decades of tracking the creature have brought interesting results.When hair that allegedly came from Wildmen was collected, the actual sources were antelope, human, wild boar, and bears.

Yeti:Asia

The Yeti is usually described as shorter than Bigfoot, and weighing 200-400 pounds.Most evidence of the creature has been what was left behind, rather than personal sightings.Individuals have seen hide, hair, and tracks.

Actual sightings often revealed the Yeti was something else.The "Yeti" ranged from a civet to a Himalayan goat.DNA samples produced similar results.The samples concluded the "Yeti" were common animals.

Yowie:Australia

The Yowie is said to be approximately the same size, with the same physical characteristics, as Bigfoot.A difference is the description of long white hair.During the 1870s, Yowie was described as having long black hair.

Started as an Aboriginal folk tale, there have been numerous reported sightings in recent decades.Although a reward was offered during the 1970s to any person who could validate a sighting by capturing one of these creatures, no one has yet claimed this reward.

The Almas -Turkish / Mongolian

There is a mountain in that country, which is thirty-two days' journey in extent. The people there, themselves say, that at the extremity of the mountain is a desert and that the said desert is the end of the earth; and in this same desert, nobody can maintain a habitation, because of snakes and wild beasts. On the same mountain, there are savages, who are not like other people, and they live there. They are covered all over the body with hair, except the hands and face, and run about like other wild beasts in the mountains, and also eat leaves and grass, and anything they can find. - EXCERPTS FROM Bondage and Travels of Johann Schiltberger

What is the Almas?

In Mongolian, it is Anmac or Almas and in Turkish it is Albis.

The Mongolian translation basically means Wild Man.These creatures are reputed to inhabit the Caucasus and Pamir Mountains of Asia and Central Asia.

The Almases are described as human-like animals that walk on two legs. Unlike the legend of the Yeti from the nearby Himalayan Mountains who is said to be massive, the Almas is said to stand between five and six and a half feet tall. Their bodies are said to be covered in a reddish bordering on brown thick hair. Facially it is said that they resemble humans with flat noses and large brow ridges with small chins. In some ways, the description is similar to that of a Neanderthal man.

The Almases have been mentioned in local legends dating back almost a thousand years.

How does this creature fit with the Bigfoot legend?

We have recorded the factual existence of subspecies of mankind such as the Neanderthal and Denisovan. Perhaps this creature was or maybe still is a subspecies of the much larger Bigfoot?

Orang Pendek:Indonesia

The name translates to short person.While accounts vary, some descriptions are consistent.It is described as a ground-dwelling, bipedal creature with powerful arms and short legs,

Barmanou:Pakistan

This creature is alternately referred to as a wild man, an ape, and a Neanderthal.Many of the sightings have been by shepherds that live in the mountainous area.

A zoologist began a search between 1987 and 1990.It was concluded that the Barmanou could indeed be a living Neanderthal.

Bukit Timah Monkey Man:Singapore

This creature is also described as a primate or a hominid.Lesser-known, with fewer sightings, reports came from residents in the area, and Japanese soldiers during World War 2.

Reports have stated the creature is gray, standing between 3 and 6 feet tall, and walks with two feet.Some believe the creatures are only crab-eating macaques.

Chuchunya:Siberia

Initially assumed to be nothing more than folklore, an expedition in 1928 concluded something really does exist.They stated the creature was similar to Alma wild people.

Natives have stated the Chuchunya is similar to a Neanderthal man, well-built, with patches of white fur on the forearms.Some stated the creature had a tail.It is said to be between 6 and 7 feet in height and wears pelts for clothing.

Grassman:Ohio

One state in the U.S.A. claims its very own Bigfoot.The Ohio Grassman is also called Orange Eyes and the Kenmore Grassman.It differs from Bigfoot and Sasquatch with its aggressive behavior.

First sighted in 1978, Grassman has been described as 5-10 feet in height, with large feet, and ranging from 300 to 1000 pounds in weight.It is loud, has a horrible odor, is exceptionally strong and very aggressive.Grassman is said to kill dogs and kill and mutilate deer.

It has a robot-like way of walking.Individuals who claim to have seen the creature stated they have seen babies, mothers, and groups.Individuals say they have seen the creatures breaking tree limbs and throwing rocks.The creature allegedly lives in nests or huts that it builds from the grass.This is the reason for its name.

Are They Myth Or Are They Reality?

Anyone who has experienced a sighting will insist these creatures are real.Scientists have not made much progress trying to prove they are not.

While there are numerous differences between creatures in different regions, the similarities they share should be noteworthy.Perhaps Bigfoot and the others are more than individuals' active imaginations.It is a thought to keep in mind if you travel to an area that has had multiple sightings of creatures that have no logical explanation.

Chapter 13: The Evolution Of Early Man

In a surprisingly short amount of time man has evolved from simple ape-like creatures into complex beings with tools, customs, civilization, culture, languages, and beliefs. Homo sapiens or modern human beings did not start with the capability to speak, make tools or even walk upright. It took millions of years for our ancestors to evolve from simple tree-dwelling primates into apex predators and the world's most dominant species. This article will delve into who these pre-human ancestors were, how they lived and evolved and how and why they died out while our species ascended to dominance.

Early Primates: Humans are evolved from early primates with monkey and ape-like qualities. Our closest living relative in terms of species are chimpanzees which share 98 percent of our genetics. Primates first began to evolve around sixty-five million years ago during the Late Cretaceous Period in the wake of the catastrophe that wiped out the dinosaurs. In the following millions of years, they began to thrive and divide into subspecies that became the forerunners of monkeys, lemurs, great apes, and man.

Humanity began to diverge from the great apes like the related chimpanzees and gorillas between four and eight million years ago and it is at this period that early hominids began to evolve in Africa.

Australopithecus: Considered by many to be the first true hominid, *Australopithecus* appeared in Africa roughly four million years ago. Unlike earlier primates, *Australopithecus* was not a knuckle-walker but rather was able to walk upright. This offered *Australopithecus* the advantage of being able to grasp low hanging fruit or spot nearby predators over long grass. The most famous specimen of *Australopithecus* was discovered in Ethiopia in 1974 and named Lucy. Lucy lived three million years ago, was bipedal and could walk upright. Lucy was much smaller than modern humans, was covered in hair like a chimpanzee had a smaller brain, did not use many tools and probably still slept in trees during the night. Her diet is believed to have been omnivorous but mostly fruit-based. *Australopithecus* represented one of the first true hominids that moved by walking on just two legs. *Australopithecus* had several subspecies such as *A. africanus* and *A. Afarensis*. It is believed that the species died out by around two million years ago, replaced by it's more. advanced decedents who continued to thrive and evolve into more complex hominids.

Homo Habilis: The next major evolutionary link in the history of man is *Homo Habilis*. *H. Habilis* lived in Africa and evolved between two and one and a half million years ago. Like *Australopithecus,* homo habilis was bipedal and able to walk upright but *homo habilis* had a much larger brain, nearly double the size. Many fossils were also found with stone tools nearby suggesting that *homo habilis* used more complex tools on a widespread basis. *Homo habilis* still resembled apes more than modern humans, with furry bodies, short stature (around four feet tall) thick brows and brains only about half the size of the average modern human's. Still, *homo habilis* marked an important transition between the fruit-eating *Australopithecus* and our next important ancestor and the species continued to exist for half a million years before finally being replaced by the more advanced *homo Erectus.*

Homo Erectus: The next major step in human evolution was *Homo Erectus*, the first hominid to stand up straight and spend its life walking or running upright. *Homo erectus* is the first hominid to appear more like a modern human than an ape, although hairier, shorter and with a flatter nose and more pronounced jaw and forehead than humans today. Homo erectus was the first hominid not to be covered in fur like an ape. *Homo erectus* were prolific tool makers and are believed to have been the first species of hominid to harness the use of fire for cooking, warmth, and light as far as a million years ago. Archaeological evidence shows they were able to fire clay and possibly have migrated using rafts. *Homo Erectus* died out around 300,000 years ago, replaced by younger, more evolved and intelligent hominids such as h*omo heidelbergensis* which then evolved further into Neanderthals and modern humans. Homo erectus was unable to keep up with these new species and were eventually replaced by them. *Homo erectus* was entirely extinct by 70,00 years ago.

Neanderthals: Perhaps the most famous of prehistoric hominids, Neanderthals lived between 450,000 and 39,000 years ago and the species shared Europe with modern humans for thousands of years. Neanderthals had a heavier frame, stronger jaw, and thicker brow than modern humans and their overall physiology were more robust than that of a modern human. They wore clothing made out of fur or hides. Neanderthals are known to have possessed knowledge of fire and stone tool making and they hunted using wooden throwing spears often tipped with sharp stones. Neanderthals had larger brains than modern human beings. They lived in groups and archaeological evidence shows that they actively cared for their wounded brethren. Neanderthal remains discovered on islands show that they were capable of building rafts or boats. It is a mystery as to whether neanderthals spoke a language or had any religious or spiritual customs. It is also unknown what caused the Neanderthals to die out but the main theories are competition or warfare with archaic humans, climate change, a disease epidemic or interbreeding with other species of hominid. Through genetic research, it has been discovered that Neanderthals and archaic humans interbred to a fairly large degree and all modern humans that do not originate from sub-Saharan Africa have at least two percent Neanderthal DNA in their genes. So while true "pure" neanderthals died out millennia ago their genes live on in modern-day humans.

Homo Sapiens Sapiens: The final link in the evolutionary chain is *homo sapiens sapiens*, otherwise known as modern human beings. As far as we know we are the only species of humans to have survived. Physically modern *homo sapiens* first appeared about 300,000 years ago but archaeology shows that we didn't start to understand abstract thought, language, spirituality or art until 50,000 years ago. *Homo sapiens* inherited tool crafting and fire-making technology from homo Erectus. Our expanding brains helped early humans to invent new technology and improve upon older inventions. Homo sapiens interbred with some of the other hominid species at the time including Neanderthals and Denisovans and so we are partially descended from these species as well. Homo sapiens spread all over the globe due to our technological prowess. A switch from foraging and hunting into agriculture and farming happened around this time and humans settled together for protection and cooperation. These settlements were the foundations of early towns, villages, and cities. *Homo sapiens* are apex predators and it is believed that our ancestors hunted several species excessively. *Homo sapiens* use of clothes, fire and constructed shelters allowed us to survive where other hominids simply could not. There is a theory that *homo sapiens* may have wiped out the neanderthals and Denisovans. *Homo sapiens* are the only hominid known to possess art, language, customs, spirituality and abstract thought. By five thousand years ago humans had created advanced civilizations, cities, myths and complex religious beliefs. Homo sapiens survived as a species by working together and using their brains to craft technologies and ideas that saved time and allowed for pursuits outside of the struggle for survival.

Are There Undiscovered Extinct Species?

Since hominids have existed for millions of years and there are several gaps in the fossil record it is entirely possible that there are multiple extinct species or subspecies of humanity that have not yet been discovered. In 2012 it was announced that a recent species of archaic hominid had been found dating as recent as eleven thousand years ago. Since prehistoric humans weren't studied until just over a century ago and excavation technology continues to advance it is likely that many more prehistoric hominids will be discovered in the future.

Could Other Species Of Humans Still Exist Today?

While there is a very slight possibility of some sort of isolated hominid species still existing this is highly unlikely as most of the world's remotest corners have been explored and it is unlikely a creature of hominid size could exist without being noticed for this long. That being said there have been other species once thought extinct that have turned up unexpectedly in the wild. This phenomenon is known as a Lazarus Taxon, the best example of which is the coelacanth, an ancient fish that was believed to be extinct for the past sixty-six million years until a specimen washed up in 1938. Since then live coelacanths have been captured, though the species is critically endangered. So while it is highly unlikely that there are undiscovered species of hominid still living somewhere in isolation it is still possible. As neanderthal DNA persists in the genes of most humans they are in a way not extinct, merely evolved into different forms.

Native American Indians and their history with bigfoot. You can expand this to include native people from around the world to include their myths and stories about the creature.

How would we track a creature like a bigfoot? How would modern technology help us to track one?

Speculate on Bigfoot's need to avoid humans is a need for any animal to avoid a dangerous predator after all man is the world's deadliest predator.They avoid man because of fear.

Here are two Chapter ideas. How do you feel about either of the two?

1.   Perhaps the reason why this creature known as Bigfoot Or Yeti or the dozen or so other names it has around the world, has never been captured or tracked is that they are highly intelligent and are good at avoiding the most dangerous of all predators on earth and that, of course, is mankind. That we cannot find them because they are afraid of us and the dangers that we pose to their species.

2. While discussing the topic of bigfoot the speculation came up that perhaps this creature is seen, but never found because it is an alien species. That bigfoot or the yeti is not native to our planet. Like UFO's these creatures are glimpsed but never anything more than that.

Chapter 14: Sighting

Bigfoot sightingS from around the world.
The historians trace the figure of Bigfoot to a combination of factors and sources including folktales surrounding the European wild man figure, the belief of folks among native Americans and loggers and the continual increase in the environmental concern. In the folklore of the native Americans, the Bigfoot is said to be hairy, upright walking, apelike creatures that dwell in the wilderness and leave footprints. Most historians often portray them as the missing link between humans and human ancestors or the great ape.

Despite the historical background of the Bigfoot, the majority of scientists have over time discredited the existence of those creatures considering it to be just folktales with no element of Truth.

Bigfoot Sightings from 1950-2019.

There had been various claims of Bigfoot Sightings, whether these claims are correct or not isn't a subject of contention. However, over time ranging from the time the existence of the Bigfoot had to be known to several journalists, historians have claimed to see some Sightings which has been recorded as Bigfoot Sightings. These Sightings would be considered exhaustively.

Most of the claims of Sightings were found in the Pacific Northwest, with the remaining reports spread throughout the rest of North America.

Thesighting is but is not limited to:

1.In November one of the 1980s, while hunting for small games an unknown reporter in Kanai- cook intet Alaska,witnessed and was opportune to get a glimpse of 'the hairy man' otherwise known as the bigfoot.

2.In Nepa country, California, in 1989, a bigfoot was sighted near the rural homes in the pope valley area.

3.In November 1968 at San Diego country, California, four camp Pendleton soldiers were privileged to see a bigfoot during their training exercise.

4.InSeptember 1970 at San Diego country California during an interview with a San Diego sheriff's sergeant, he said there had been a series of incidents of bigfoot sightings after the laguna fire incidents of 1970.

5.In rounding off this article in June 1993 at Tuolumne country in the state of California, a wildlife biologist observed a large upright walking animal near Pinecrest. He observed that; in the word of the biologist he gave the account of his experience he said; The following is an account of the events of the day while I was conducting surveys for amphibians downstream of Bloomer Lake. The account may be somewhat fragmented in presentation, as I want to provide a narrative that portrays the events of the day. Additionally, I prefer to provide a narrative and let people draw their own conclusions first, I recall finding the remains of a fawn in the meadow I was surveying. Nothing unusual about that, in-and-of-itself...ties-in later. As I continued to conduct my survey, perhaps thirty minutes or so afterwards, I noticed movement off to my right. What I saw was what appeared to be an animal covered with black, semi-long hair (two to three inches, to guess), walking upright (fully) with a gape somewhat like the current hominids; and approximately five to six feet in height and weighing one hundred seventy-five to two hundred twenty-five pounds. I saw something move for two or three steps. It did not appear to go anywhere (but just moved) but of sight. It did not look at me or react to my viewing, it just simply appeared and then was gone.

To some extent I was not really paying attention to my surroundings, as I was focused on my work; but was enjoying the surroundings in that field sort of way. I recall thinking that that looked like an upright creature walking through the woods. I tried to repeat my last movements, to see if what I saw was a play on light and shadows. I was not able to re-create any likeness of what I had seen - the light and shadow of the forest did not seem to be the source of my observation.

I did not hear anything. Either before, during, or after, in regards to the sounds that one might hear if someone were walking in the woods stepping on branches or simply rustling the ground. I did not take an active interest or pursuit, as I was uncertain of what I think I really saw; I am skeptical by constitution; not only for mythologies, but in general.

I did play the event in my mind while I finished my survey. And occasionally looked in the general vicinity of the area of where I saw the animal-creature-mythological being. I finished my survey and left the area.

While I was returning upstream, approximately one hundred meters from the meadow, I came across an exhausted buck - two or three-year-old lying on the ground near the creek. I was able to approach and touch the deer. I clearly recall brushing flies off his head and sitting next to him for a couple of minutes and thinking this is pretty cool! I was curious about the deer and did think about what I had seen in the meadow.

I then proceeded upstream again. Approximately one hundred fifty meters from the buck I came across a mass of tissue, which looked to me like an aborted fetus. Upon further investigation, I discovered that it was the entrails of an animal. Very fresh. I then looked behind me and saw a freshly killed doe. Her eyes were still clear and there were no flies or maggots on or in the carcass. As I examined the doe I could find no sign of struggle. I was unable to locate any apparent method of kill by another animal; thinking of a cougar, which usually will bite at the base of the skull or that general area. I found no such evidence. The carcass was left fully in the open. What I did notice was that the right leg had been cleanly ripped off from the main body to include a couple of ribs. There was no blood anywhere. I looked around as my in-the-woods instincts were quite active. There was that six sense of a presence. I spent a while running all of the pieces through my head and thought that this was pretty cool and funky. I then continued to hike out to Bloomer Lake, taking the experience with me.

In conclusion, it is true to say that there had been several occasions of Sightings of Bigfoot although the existence of Bigfoot had not been scientifically proven, however subject to folktales and folklores there had been and in fact, there are still Bigfoot Sightings.

Chapter 15: If Bigfoot Exist

He really doesn't but...

What if he did. There would be a lot of questions that need to be answered as the grainy film footage does not do the beast justice. What will take place in this chapter is an attempt to answer those questions.

What would we call him...or her

The names Bigfoot, Sasquatch, Yeti, and Yowie are more modern monikers and may not be applicable. Maybe the beast would be insulted by being named and remembered for only the large size of his or her foot.

There are more historical names applied to the creature and these names are given to the creature by native American Indians who have many legends about their encounters with the beast.

These people groups used names like Ts'emekwes, Stiyaha, or kwi-Kwiyai. Deciding on a name would be difficult because it seems that this creature goes by so many.

What would Bigfoot look like

This is a very good question because the only images we have are usually out of focus, grainy photos, and videos. These images only depict a single member of the Bigfoot species. There are no photos or pictures of the whole group standing together, bathing, or fishing.

What if the star of those images was the Big Foot's Andre the Giant? He or she, was not the actual size of the species but was an anomaly? The baby had a thyroid or growth hormone condition that brought him, or her, to a very large size and the rest of the species was only 4 feet tall?

It is hard to tell what the rest of the species looks like because we only get one representative at a time. And if that representative was the female of the species, the world may shudder to think what the male looked like.

Also, would the rest of the species look like a bear? An African Gorilla? It is hard to say because there seems to be only 1 star of all those Bigfoot captured on film pieces of evidence.

Where do they live

This is another big question because the species would need to have a home somewhere. No self-respecting male would have his bride or his children, live in the forest unprotected and vulnerable to all the hikers, campers, adventurers and even Bigfoot hunters to harm them.

If they live in caves, where are they? It is kind of certain that the large caves and caverns that would be needed to house these animals would be stumbled upon by these wood loving humans.

They must be very smart creatures to be able to hide their homes from so many searchers for such a long time.

Where are their trash and waste?

If Bigfoot exists, these creatures display an enormous amount of societal intelligence and planning. In the hundred years or more of searching for this species not one sign has been left that would indicate where the creature deposited their waste and their leftover meals.

Not one. That lack of discovery makes the Bigfoot smarter than most creatures, perhaps even man. They are also very environmentally friendly as their activities do not harm the environment in any way.

Would we recognize a Bigfoot...

If one of the many Bigfoot hunters came across a dead Bigfoot's body? Especially if that body did not look like the sole representative that has been captured in so many bad, blurry images.

This also is an interesting and important question for we would not be able to prove that Bigfoot existed if we can't identify the body. What if the body was a lot smaller than the one everyone sees in those grainy videos, what then?

Would it be identified as Bigfoot or an escaped orangutan from a traveling circus or nearby zoo? What evidence would be used to make the identification seeing that no one actually has any evidence that would aid in that process?

It would be fun to watch the scientists argue with each other over the identity of the dead body. Odds are if the inner works of the creature were unusual, there would be a horde of 'scientists' claiming it was an alien.

If Bigfoot existed

It would be an amazing creature. Its intelligence levels must be off the charts since it has escaped discovery longer than any human has been able to do. It might be the gentlest of creatures who follow the example of Greta Garbo who once said 'I want to be left alone'.

What it looks like may scare a few people or it just may be as cute as a baby chimp. It is hard to say. It is also hard to prove it actually exists as it covers its tracks quite well.

Chapter 16: Is Bigfoot Highly Intelligent

Bigfoot – A Higher Intelligence?

Across the continental U.S, thousands of sightings are reported each year of hairy, ape-like creatures otherwise known as bigfoot, with reports either unverified or unreliable. In 2019 alone, 23,000 sightings were reported with Washington leading the ranks with the most documented appearances; almost 2,000, according to the Bigfoot Field Researchers Organization. However, with the increase in sightings and the rise of modern technology, the question becomes not if the creature exists but why evidence remains in grainy footage and mysterious stories. The evasive nature of the creature not only tells of higher intelligence but a creature aware of the dangers of humankind. In each report, similar elements are relayed. First, of aggressive warning behaviors, rock-throwing, and tree hitting, followed by the creature seamlessly moving through woodland, leaving little to no evidence. A hiker walking through the Provo Canyon in

2012, recounted a similar experience crossing paths with a bigfoot. When after seeing the creature crouched behind a tree, he was shocked when it stood on two legs and threw a large rock towards him as if warding him out of the area. Another encounter occurred in 2013 when a hunter in Canada reported a bigfoot while seal hunting. Spotting what looked like an animal on a ridge, the hunter and his group followed the creature until they heard rocks being thrown through the trees. The group speculated the creature was angry and warding them out of the area since they'd interrupted its hunt for food, most likely caribou.

Animals respond to threats with an acute stress response, namely a fight or flight reaction, which can manifest in different complex reactions. For example, the myotonic goat, which falls over when startled. However, in threatening situations, bigfoot creatures react in calculated methods, yelling between familial parties, hitting trees, and throwing objects until the threat has departed.

The behavior is similar to intelligent primates, which like bigfoot, communicate for cooperation, and compete for resources. In this case, it's arguable the competition for resources is between bigfoot and humankind, and the resulting consumption of society. The global tree cover is steadily declining, with an estimated 15 billion trees cut down every year for goods. Though the United States still has around 200 billion trees across the country, the loss of biodiversity could be a contributing factor to the evasive and stealthy measures of bigfoot and its increasing awareness of humankind. Going back to the first sightings in the 1800s of people reporting a 'wild man', bigfoot has been an illusory figure just out of reach and may have recognized humankind as a negative factor in their environment from the very beginning. In 1840, a missionary by the name of Elkanah Walker relayed the stories from the Indigenous Americans regarding giants living in the nearby mountains in Washington, benign creatures that stole fish from the locals. This countered other tales across the country later in the 1800s of menacing wild men carrying people off and killing them, which suggests the current aggression could be a result of unanticipated land loss and competition for dwindling resources. From then on, sightings increased and particularly exploded in the 1950s when a small article was published on the hairy creature in the *Humboldt Times* comparing the beast to the abominable snowman. All over the world, people reported seeing an 'ape-like' man with names such as the Yeti, Almas, Batutut, and the Yowie giving recounts of their stories. However, the evidence leading up to the present day is sparse. The most convincing evidence is relegated to blurred camera footage, plaster cast footprints, and traces of DNA left at the scene in hair samples. In 2012, one such specimen analyzed by scientists led to the announcement of bigfoot being an unknown primate species and a distant human relative known as Homo sapiens cognatus. Though the publication of the announcement was met with contention, the deceptive and intelligent nature of the creature avoiding humankind could support a long lost primate connection. Other explanations involve extinct hominins traveling from Asia during the ice age to northern America, with a rounded skull capable of supporting a large brain for cognitive capacity. The origin might explain the intelligent methods known for evading capture, such as rock-throwing and intimidation, similar to early human beings and primates, particularly chimpanzees. Bigfoot organizations relay similar anecdotal evidence of intelligence with the creatures having family structures and other behaviors, for example, constructing shelters to sleep in with ferns and other shrubs used for padding. Bigfoot vocalizations are equally complex and range from whistling, howls, grunts, or roars to communicate to family members who more often than not move in groups. Other evasive methods are credited to the creature's ability to move without leaving evidence. They're reported to glide through the trees, almost effortlessly, with long strides propelled on bent legs, leaving little to no trace. In nearly every encounter, the creature is nothing but a blur among the trees being able to run almost 40 mph. But in some reports the creature withdraws from danger, retreating out of harm's way unless scaring people from their territory, similar to great apes, by chest-thumping, yelling and breaking branches.

According to Bigfootfinder.com, bigfoot are territorial creatures, purposefully choosing remote areas away from people where there are few threats and plenty of food. The fact that their habitat is under threat, only further suggests that the aggression is pointed to humankind as being the danger to their species. But the evidence so far is only the start of understating bigfoot's ability to evade capture. For over 60 years, reports are unreliable at best, which could suggest an intelligence greater than hypothesized. What is the possibility of a tall creature populating wilderness areas without leaving evidence or confirmation of existence? No evidence of leftover meals, bodies or habitation,besides makeshift shelters. Intelligence can only be gauged by the evidence provided, and perhaps the intelligence of bigfoot is more significant than realized, possibly even surpassing our own. Evading technology, human interference, and hunters, bigfoot's intelligence is unmitigated and as evasive as its capture.

Chapter 17:Alien Or Dimensional Visitor

Bigfoot – Interdimensional Visitors

Since the early settlement of the United States, bigfoot sightings have been common phenomena, followed by UFOs and other unexplainable land creatures. But a new theory challenges humankind's understanding of biology by a connection between these strange occurrences; that bigfoot is neither an ape or a long-lost human relative but an extraterrestrial being unnatural to our world. UFO and bigfoot finding communities alike dismissed the connection until the 1970s when eyewitness testimony accounted for a different story.

Starting from the late 1800s, hairy, humanoid beings have been spotted alongside unexplainable light events that appear in the sky in strange patterns. A cattleman from Humboldt County, California, first reported hearing of the beings when he spoke to the Native American locals. The locals described what they called 'bear-like' creatures, which had been dropped from the moon and sat in nearby caves. Today, similar sightings are reported with almost 20 percent of bigfoot incidents occurring alongside light phenomena. John Keel, the author of *The Mothman Prophesies*, hypothesized the events were related, and that bigfoot creatures were ultraterrestrials, alien beings capable of manipulating time and matter. According to Keel, these beings can materialize and dematerialize in our dimension, which could explain the lack of evidence in finding the creatures. Geographical 'window areas' known for high magnetic fields are said to be required for materialization, such as Washington, Oregon, and Pennsylvania. In Pennsylvania, a high number of bigfoot/UFO sightings were reported in 2019, but the most memorable encounters occurred in the 1960s. One such event was reported near Lake Erie when a group of tourists got their car stuck on the lake's peninsula. Around 10 pm, two police officers stopped by the stranded tourists who said they witnessed something strange happening in the woods. While the officers and two of the group members went to investigate, a black being, with no hands or face, was spotted on the beach, followed by an orange light shining in the woods. Triangular marks left at the scene caused the officers to speculate where the craft had landed, though the CIA later dismissed the reports as a hoax, even as multiple witnesses claimed to see the event. Hundreds of sightings throughout the 1970s continued to establish the connection between bigfoot and extraterrestrials as more lights and humanoids were spotted. In 1973, a woman and her 13-year-old son were sleeping in a trailer in Cincinnati, Ohio, when unexplainable cones of light appeared in a nearby parking lot. Minutes later, the woman saw a gray, hairy creature which then disappeared with the lights. If bigfoot creatures are alien beings shifting through time and space, the theory would support the lack of evidence since the first appearances in the 1800s. The evidence so far gathered is restricted to unverified camera footage, plaster cast footprints, and small hair samples, which has only obscured the presence of the creature further. According to paranormal research, only an intelligent being capable of technical sophistication would be able to avoid detection by human radar and other modern technology. With a lack of noticeable habitation, carcasses, food scraps, and other droppings, the creatures seem to vanish into thin air. However, other theories have been posited to the appearance of the hairy ape-like creatures and lights. One such theory states that the bigfoot creatures may play a similar role as cats and dogs, being pets for their alien owners, the appearance of the lights really being extra-terrestrials abandoning their pets. In other theories, these creatures, like humans, are abducted for extra-terrestrial experiments related to gathering information about our species. But the lack of evidence undermines the theory. Paranormal researcher, William Hall, states that the paranormal world may be connected in ways previously unthought. In events with paranormal beings, there are similar signs, lights, creatures, objects, which are dismissed as coincidences. As Hall states, this is due to the disconnect between UFO and the bigfoot finding communities. Up until the 1970s, the groups were reluctant to acknowledge the other due to speculations, which at the time seemed coincidental rather than a connected phenomenon. The clustered 'window areas' throughout the United States corroborate the alien/bigfoot theory as sightings have been plotted in similar areas. Peter Leeson, a professor at George Mason University, began researching UFO and bigfoot sightings and reported strange geographic patterns in states such as Colorado and Alaska, which confirm these hotspot areas. But are the creatures really aliens crossing into our dimension or alien pets abandoned by their owners? John Keel proposed another concept that the events were merely hallucinatory: "Some UFOs were directly related to human consciousness, just as ghostly apparitions are often the product of the percipient's mind." Dr. Persinger of Laurentian University states something similar that the events are hallucinatory experiences caused by the brain's response to magnetic energy in some geographic regions; the unexplainable lights and hairy creatures a by-product of the brain's processes. Similar to the bigfoot/alien theory, it substantiates the resultant lack of evidence for the creature but doesn't explain the up-close sightings. In these encounters, witnesses dodge huge rocks thrown through the woods by bigfoot as a warning signal, followed by howls, screeches, and ominous knocking on trees. In one instance, a witness reported ape-creature breaking trees and even mimicking the sound of earth-moving equipment used on their property. With encounters so tangible in nature, the likelihood of a hallucinatory experience becomes less likely, as a large number of people sometimes experience such events.

With each theory proposed by experts in the paranormal field, the chances of an interdimensional connection with alien beings become stronger and a shared ground for enthusiasts. Either an alien, an alien pet, or a hallucinatory experience, the evidence displays strange occurrences in our world, something beyond understanding. Are we really alone or victims of interdimensional beings? As long as the sightings increase in geographic hotspots, humankind can never be sure. Our world is open to those with the capability to travel through it.

Chapter 18: What Is Cryptozoology

Cryptozoology is considered a pseudoscience that studies the existence of mythical creatures found in folklore and legends. There has been some form of cryptozoology since early civilizations. Today stories of these mythical creatures, also known as cryptids, appear in a variety of cultures across the globe.North America has Bigfoot, Mexico has Chupacabra and Scotland has the Loch Ness Monster. Although the academic world does not recognize cryptozoology as a true science, mostly because it does not use the scientific method, it's history is intriguing and has at least brought attention to how humans process things they do not understand.

Bernard Heuvelmans and Ivan T. Sanderson are credited with pushing the ideas of cryptozoology into the public's eye in the 1950s. Heuvelmans' book On the Track of Unknown Animals (1955) and Sanderson's book Abominable Snowman: Legend Come to Life (1961) lead to an increase of attention to the pseudoscience and other works about the topic being published. Cryptozoology had an audience. By 1982 Heuvelmans created the International Society of Cryptozoology and was printing a journal. In the journal is where the term cryptid was first used and is not recognized as a term by actual zoologists. The group lasted until 1998 and had to stop their research due to financial problems.

ICS used the okapi as its symbol. An okapi is an animal related to a giraffe, has markings that resemble a zebra and resides in the Congo. Harry Johnston, an English explorer, introduced the okapi to the scientific community in 1901 and it received a great deal of attention. Instances like these would be the most valuable idea that ICS would bring to the scientific world. When species like the duck-billed platypus are discovered it bewilders the science community and many theories about how this animal came about are created. Throughout history, there are reports of a variety of monsters like animals that turned out to be real. The Komodo Dragon, the manatee, and even the gorilla were all considered monster myths until someone came along and used science.

After the ICS disbanded a group following the ideas ofYoung-Earth creationism became the face of cryptozoology. Although the YEC was already considered a subset to the pseudoscience, paleontologist Donald Prothero said in 2013 that creationist have become the strongest group pushing the ideas of cryptozoology. He claims their intentions are to disprove the existence of evolution by discovering species that show inconsistencies in the ideas promoted by evolution. "They think that if they find a dinosaur in the Congo it would overturn all of evolution. It wouldn't. It would just be a late occurring dinosaur, but that's their mistaken notion of evolution," Prothero told Racheal Shea, a reporter for the National Geographic, in September of 2013.

Most other paleontologists align with Prothero's view on cryptozoology. Paleontologist George Gaylord Simpson refers to the pseudoscience as an example of human gullibility and compares it to those who believe in alien abductions. Peter Denzel, a folklorist, says cryptozoologists purposely defy accepted mainstream science by not using any true science to prove their case. Many have also noted that there is not a school that carries a cryptozoology degree and many of the people who involve themselves in this study do not even have a science background. Those that do have academic backgrounds do not have a background that would prepare them to research exotic animals.

Not following the scientific method and being associated with fanaticism are not the only problems that cryptozoologists face when trying to explain their reasons for chasing mythical creatures. The media helps out a lot. Newspapers like The

National Enquirer or The Onion print stories that are obviously false information meant for a laugh. Television shows about mysterious monsters and movies that present the creatures as supernatural, sometimes in a very comical way, add to the uphill battle for those who wish to prove their existence.

The biggest issue facing the credibility of cryptozoology is a result of those who profess its validity. Many of the claims or sightings that cryptozoologists have presented were easily rebuked by real science or the person who made the claim admitted it was a hoax. The Loch Ness Monster is a classic example. The pictures were proven to be fake, sonar showed there were no whale-sized creatures in the loch and more recently a sample of the water showed an extreme amount of eel DNA, one of the lake animals thought to be mistaken as the monster.

The creation of misinformation from fear of the unknown is common throughout human history. Crop circles, the pyramids, ghosts, and mythical creatures grab our attention because of the inability to explain their existence. Through science, many of these occurrences have been explained, but the ones that have not grabbed the attention of people who want to believe that there is more to existence on this planet than we know. Attempts to explain the unexplainable as acts of gods or aliens from far beyond will probably continue. Is there a giant octopus in the sea or a terrible monster that resembles a gargoyle terrorizing New Jersey?

Cryptozoology, like many other pseudosciences, will probably be around for a long time to come. Whether or not you choose to believe in the ideas of ICE, or the creationist trying to disprove evolution, it can not be denied that folklore and the history surrounding these creatures does seem a bit coincidental. When multiple descriptions given by people who claim to have seen a mythical creature continuously sound similar one might want to research and figure out its validity.

For most, cryptozoology will just be for entertainment. Watching science fictional films and letting our imaginations take control for a while is always a great way of stepping back from true and many times stressful situations that reality presents. Maybe one day a Sasquatch, Jersey Devil, Yeti or a giant octopus is known to wreak havoc in the seas and sink giant ships will be caught and all the cryptozoologists can have their moment. Until then enjoy the myths and excitement of the unknown on occasion sounds a bit more rational.

Chapter 19: Similarities Between

Sightings

Yeti; also known as the **Abominable Snowman** is a folkloric ape-like creature taller than an average human, that is said to inhabit the Himalayan mountains. The names *Yeti* and *Meh-Teh* are commonly used by the people indigenous to the region and are part of their history and mythology. Stories of the Yeti first emerged as a facet of Western popular culture in the 19th century. The scientific community has generally regarded the Yeti as a legend, given the lack of evidence of its existence just like the Bigfoot, the Scientists do not believe in the existence of Yeti they are taken as a historical myth.

The historical story of the Yeti dates back to the early 19th century. And the formation of the name Abominable Snowman was in the year 1921 the same year Lieutenant-Colonel Charles Howard-Bury led the 1921 British Mount Everest reconnaissance expedition which he chronicled in *Mount Everest The Reconnaissance, 1921*. In the book, Howard-Bury includes an account of crossing the Lhakpa La at 21,000 ft (6,400 m) where he found footprints that he believed "were probably caused by a large 'loping' grey wolf, which in the soft snow formed double tracks rather like those of a bare-footed man". He adds that his Sherpa guides "at once volunteered that the tracks must be that of 'The Wild Man of the Snows', to which they gave the name 'metoh-kangmi'" "Metoh" translates as "man-bear" and "Kang-mi" translates as "snowman".

Features of the Yeti

The Yeti is said to be muscular, covered with dark grayish or reddish-brown hair, and weigh between 200 and 400 lbs. (91 to 181 kilograms) It is relatively short compared to North America's Bigfoot, averaging about 6 feet (1.8 meters) in height. Though this is the most common form, reported Yetis have come in a variety of shapes.

The existence of yeti.

Just like the bigfoot, the scientists do not believe in the existence of Yeti, however, it can be said to be folklore, the existence of yeti can, however, be proved by its numerous sighting which includes;

·In 2007, American TV show host Josh Gates claimed he found three mysterious footprints in the snow near a stream in the Himalayas. Locals were skeptical, suggesting that Gates — who had only been in the area for about a week — simply misinterpreted a bear track. Nothing more was learned about what made the print, and the track can now be found not in a natural history museum but instead in a small display at Walt Disney World.

·In 2010, hunters in China caught a strange animal that they claimed was a Yeti. This mysterious, hairless, four-legged animal was initially described as having features resembling a bear but was finally identified as a civet, a small cat-like animal that had lost its hair from disease.

Drawing inferences from the sighting it appears that unlike the bigfoot, the existence of Yeti had not been scientifically proven.

Distinct Difference Between The BigFoot And Yeti

The difference between these two creatures is in their location;

·Yeti is found in the Himalayas and has its roots in pre-Buddhist religion. While the Yeti belongs to Asia, Bigfoot is thought to be native to North America, specifically the Pacific Northwest.

Similarities between the yeti and bigfoot:

·Despite originating thousands of miles apart, some modern-day believers suspect that the creatures belong to one species. One popular theory is that Bigfoot and the

Abominable Snowman/Yeti are both *Gigantopithecus*, a polar bear-sized ape native to southern Asia believed to have gone extinct 300,000 years ago. While chances are slim that the species migrated to North America with its *homo sapiens* relatives, that hasn't stopped many cryptozoology enthusiasts from wanting to believe.

·The two mythical beasts look like each other. All of them look like big, really big (7 feet or more) humanoid hairy bipeds walking on their foot, looking like giant ape-men, leaving giant footprints looking like an oversized human step.

Although the existence of these creatures has not been scientifically proven they can be said to exist due to the historical folktales and their various sightings.

Chapter 20: A True Encounter From India

Hello guys, I am Aditi. I am from India and I am a very adventure-loving person. I mean, who wouldn't love to have a little fun and danger at the same time, right? I have always been that kind of person who would go to some scary haunted house or some forbidden deep forest areas just for fun. In short, I want to go everywhere I am not supposed to go. And this trait of mine is equally admired and feared by my friends. You know, addiction to danger is bound to put you in it someday. And this was my turn.

It was late August. My birthday was coming. And as usual, my friends Disha and Simon and my boyfriend Akash refused to do something usual for me. They surprised me every year with something simultaneously amazing and stupid. So, this year it was trekking. And not any common route, they found out the most mysterious route for us to explore. And not to mention, I was beyond excited. We planned to blow the candles in the Miyar Valley which was one of the most dangerous trekking spots located in the

173

Himalayan mountain range. So, as my birthday was only 8 days away, we packed our things and got on the train to Chandigarh. It took us around 3 days to reach Chandigarh. It was afternoon when we reached there and we were supposed to start on the next day.

The next day we drove to Manali. It was a long drive and we reached there in the evening. We planned to head to Urgos the next day and finally reach to Khanjar the day after from where we would be able to start the fun part, CLIMBING! So, the next two days were almost spent in the car and finally two days before my birthday, in the morning we started climbing our way to Miyar Valley. Like every other trekking trip, the first few hours were all fun and games. We were all used to trekking and that is why none of us needed any rest for the first few hours. We were chatting and climbing and clicking photos. The estimated time to reach the top of the valley was two days. We were carrying tents in our backpacks along with other essentials.

The Miyar Valley has an altitude of around 18,000 feet. We kept climbing through the snow and it was quite exhausting after a few hours. So, we decided to put up the tents when the lights were almost gone from the skies. The night falls real fast in the mountains and we were lucky to find a spot beside a brook halfway to the top. It was a beautiful and quiet place. The four of us were exhausted beyond measure by that point. We weren't any less enthusiastic though. We had no hope of finding dry wood for a fire in that forest filled with snow. We set up the tents on a comparatively usable spot. I and Disha had a tent for ourselves and Akash and Simon put one up for themselves. We were carrying dry food and water. We were hungry was an understatement. We were LITERALLY STARVING! We ate and talked for a few minutes around the solar-charged light we were carrying to scare the wild animals away. Then we headed to our respective tents to sleep.

While I was sleeping, I heard something. I remember wondering, half asleep, how exhaustion results in snoring because I was sure that it was Disha snoring beside me. But man, that sound was loud! I was about to roll to my side to ask her to stop when suddenly I felt a shake on my hand. "Do you hear that, Adi?", Disha asked, panicking. I heard what sounded like a low growl at first. "I think it is some kind of a wild animal", I said uncertainly. This part of the forest was supposed to be deserted even of wildlife. The area was really cold and almost inhabitable due to the lack of trees at this height. The growl sounded as if it was moving at a distance from where we were. The boys were also up by that point and they were entering our tent moments after we woke up.

Akash was the bravest of us all. He seemed quite curious about what the source of the sound was because the more we listened to it the weirder it got. It was not exactly the call of some wild animal. We were all pretty sure about that because we have camped a lot and we never heard some sound like that. Sleep was far away from us now and we concentrated on the sound. Although it sounded like a growl at first, by listening carefully it sounded more like footsteps, huge footsteps.

We were scared. It was a really strange feeling, sitting there, four people in one small tent, and listening to that sound. We had no idea what it was. Akash and I wanted to go out there and have a look but the other two stopped us. At last, we also agreed that it would be quite foolish to go there if it was indeed some wild animal. But it felt like something else. I had this feeling in my gut that it was something more than just some wild animal. I wanted to know what it was. I wanted to see if what I was thinking was true or not. I heard the stories. I believed them to be myths but then again, just because we haven't seen something doesn't mean it wasn't there. I have always been a believer. And I knew that my friends were sitting there thinking the same things.

We sat there like that for a long time. None of us spoke. We listened carefully. The sound seemed to travel parallel to the path we were trekking on. As if it was heading to the top of the mountain as well. It was getting closer at first but then it started to fade slowly. I have no idea how long we were sitting there like that actually. But it seemed like an hour had passed when we were finally unable to hear that sound anymore. It was 3 in the morning and none of us had even a bit of sleep in their eyes. We started voicing our opinions on what that was and just as I thought, we were all on the same page.

When the sun finally came up, we decided that we needed answers. We were sure that the sound was coming from the east. So instead of heading north according to our plan, we headed east. We were sure that it wasn't very far from where we were because the sound was pretty clear. It was easier to go sideways on a trek instead of climbing up. We were going quite fast. Then suddenly, something very strange caught my eyes. Holes. Gigantic holes in the snow. I called out for my friends who were a bit behind me and they came running to the spot where I was standing, frozen. The snow was pretty deep in that area. And in the deep snow were holes, no not holes, footprints, almost ten times larger than mine. They seemed to be heading upwards and I had no idea what they were of. It couldn't be of elephants because they can't climb those steep pathways.

I have no idea what it was. But one thing we all were sure of. We did not want to meet that. Loving adventures and risking lives are two completely different things and we were not down for the latter. We got back to the foot of the mountain as soon as we could and I celebrated my birthday on sea-level. I still can't surely state that what we witnessed was Bigfoot or Yeti, but after that day I know that I believe in them. What I witnessed made me believe that. It could not be anything else after all.

Conclusion

Now you know about some of the most astonishing sightings, encounters and evidence of the world's many mysterious ape-like creatures. By examining the eyewitness descriptions, it is clear that some of them, like the Yeti, Yowie, and Bigfoot, seem very similar, if not identical, while others, like the Almas and Orang Pendek, seem to stand out from the crowd.

However, if we assume that the beings are, in fact, real, all of them certainly share an exceptional ability to stay hidden from humans. Untold numbers of expeditions, both by independent actors and big organizations, have been launched in an effort to find Bigfoot and the others. They have scoured the deepest reaches of the Earth's forests and the tallest peaks of faraway mountains.

Still, even though some have returned with compelling evidence, such as casts of footprints, hair samples or even semi-decent video footage, nobody has been able to fully prove the existence of these beings. But how could that be? If the eyewitness reports are real, and large, ape-like creatures live in remote areas around the globe, we would except to find something more tangible — like a dwelling place or the remains of a dead specimen.

While most cryptid enthusiasts camp out in mountainous forests hoping to catch a glimpse of the strange creature, one possibility seems to have eluded the majority: Bigfoot and the others may live most of their lives underground. Sure, the hairy hominids seem to search forests and mountainsides for food and other resources, but there are many clues that point towards them being mostly subterranean.

For example, consider the Native American tales of Sasquatch that go back long before European settlers even imagined the creature's existence. Some tribes, like the Mono Peoples of Sierra Nevada, California, consider the creatures to be fellow inhabitants of the land. Through generational teachings, they understand that the intelligent beast-men live in the caves and tunnels that run beneath the Sierra region, and only come out on rare occasions.

The idea of these creatures being largely subterranean also fits nicely with the reports of them being mostly nocturnal. Unlike humans, they seem to have no trouble seeing and moving about at night. Eyes like that would most likely be well-fitted for underground living as well. Combine all of this with the fact that most Bigfoot sightings have taken place in, or around, a mountainous area, and this theory seems to make a lot of sense.

With this in mind, perhaps cryptozoologists would be well-advised to examine not only remote forests and mountain ranges, but also the vast, unexplored caves and tunnels that lie underneath them. While this would certainly require more effort and additional safety measures, who knows what could be found in the deeper places of the Earth?

One thing is for certain, however: sightings of Bigfoot and the others are still being reported to this day. With ever-improving, easily-accessible technology, hopes are high that we will eventually get to know the elusive ape-men of our world.